Artificial Intelligence IN AFRICA

By Doye T. Agama
With claude ai

Artificial Intelligence in Africa

Artificial Intelligence in Africa

Copyright © (Doye T. Agama 2023)
All Rights Reserved

No part of this book may be reproduced in any form by photocopying or any electronic or mechanical means including information storage or retrieval systems, without permission in writing from both the copyright owner and publisher of the book.

ISBN: 9781804678671
Perfect Bound

First published in 2023 by bookvault Publishing, Peterborough, United Kingdom

An Environmentally friendly book printed and bound in England by bookvault, powered by printondemand-worldwide

Contents

Contents

- INTRODUCTION 7
  - The Fourth Industrial Revolution 7
  - Why is this book for you? 9
- CHAPTER 1 - WHAT IS AI? 12
  - The Core Idea in Chapter 1: 12
  - AI Explained 12
  - Big Data 13
  - Meet the Continent of Africa 15
  - AI - An African Opportunity 20
- CHAPTER 2 – AI INVESTMENT IN AFRICA 22
  - The Core Ideas in Chapter 2 22
  - Africa's AI Investors 22
  - Strive Masiyiwa 23
  - The African Development Bank 25
  - UNICEF, Google, Microsoft & more… 28
  - Growth in African AI Investment 33
- CHAPTER 3 - AI CASE STUDIES IN KEY SECTORS 36
  - The Core Ideas in Chapter 3: 36
  - Healthcare - Predicting Disease Outbreaks 36
  - AI For Job Creation 40
  - Multi-Dimensional Interventions 41

CHAPTER 4 - DATA PROTECTION AND AI GOVERNANCE 44

    The Core Idea in Chapter 4: 44

    Human Development and Political Instability 44

    Political Uses of AI 46

    Data Governance in Africa 49

    Data Protection Africa 49

    AI Data Bias 51

    Data Rights in Africa 53

    Data Protection Laws in Northern Africa 56

    The African Union 58

CHAPTER 5 - DEVELOPING AI TALENT AND SKILLS 61

    The Core Idea in Chapter 5: 61

    African and International Skills Gaps 61

    African AI Skills Growth 63

    African Based Degrees in AI 63

    Egyptian Startups Accelerator Programme 68

    African Robotics Network (AFRON) 69

    Pan-African Robotics Competition 69

CHAPTER 6- WHAT NEXT? 71

    The Core Ideas in Chapter 6: 71

    The Micro Entrepreneurs 72

    M-Kopa in Kenya 74

    THINK BIG! 75

Starting Your Own AI Business 76
Machine Learning Project Readiness Tool................................... 76
Emerging AI Apps 78
Key skills for AI/Machine Learning Entrepreneurs: 80
Funding Application Criteria 82
CHAPTER 7 - CONCLUSIONS 84
Regional Cooperation... 88
End Note 91
Suggested Further Reading: 94
GLOSSARY OF TERMS 95

INTRODUCTION

The Fourth Industrial Revolution

The First Industrial Revolution of 1765 to early 1800s was characterised by mechanisation that changed the major industries of the time such as coal mining. Africa participated by providing raw material and the "free" labour of enslaved people.

The Second Industrial Revolution of 1870 to the middle of the 20th Century was the harnessing of new energy sources like oil and gas, the invention of cars and new aircraft, development of modern industrial chemistry and the beginning of the telecommunications revolution. The participation of a largely colonised Africa was again peripheral.

The Third Industrial Revolution 1969 till 2000 sees the growth of nuclear power plants, massive scaling of telecommunications and revolutionary changes in electronics that make new levels automation and robotics possible[1]. By now we see the impact of waves of political independence across much of Africa, hampered by neo-colonial (and Cold War) apron strings. But these years coincide with an explosion in education and other opportunities which slowly begin the change in African self-identity.

The Fourth Industrial Revolution (Industry 4.0 or 4IR) is a fairly new expression coined or popularised by Klaus Schwab, founder and executive of the World Economic Forum in 2016[2]. Industry 4.0 which started in the early 2000s is till happening all around us.

---

[1] https://ied.eu/project-updates/the-4-industrial-revolutions/ (Accessed 25/09/2023)
[2] Philbeck, Thomas; Davis, Nicholas (2018). "The Fourth Industrial Revolution". Journal of International Affairs. 72 (1): 17–22. ISSN 0022-197X. JSTOR 26588339

According to the Brookings Institute, it has huge opportunities for Africa where technology can impact the economies and improve the lives of people. But there is also the potential for inequalities to widen as part of possible disruption caused by the new technologies[3].

In the Salesforce 360 Blog "What Is the Fourth Industrial Revolution?" Devon (Pasquariello) McGinnis, Director of Content of Editorial Strategy at Salesforce, describes The Fourth Industrial Revolution as, *"a way of describing the blurring of boundaries between the physical, digital, and biological worlds. It's a fusion of advances in artificial intelligence (AI), robotics, the Internet of Things (IoT), Web3, blockchain, 3D printing, genetic engineering, quantum computing, and other technologies. It's the collective force behind many products and services that are fast becoming indispensable to modern life. Think GPS systems that suggest the fastest route to a destination, voice-activated virtual assistants such as Apple's Siri, personalized Netflix recommendations, and Facebook's ability to recognize your face and tag you in a friend's photo"*.

Devon goes on to argue that unprecedented change and disruption in every aspect of our lives is a result of this "perfect storm of technologies," and shows how her company has responded by building new products to meet customer needs[4].

---

[3] https://www.brookings.edu/wp- content/uploads/2022/03/4IR-and-Jobs_March-2022_Final.docx.pdf

[4] https://www.salesforce.com/blog/what-is-the-fourth-industrial-revolution-4ir/ (Accessed 25/09/2023)

Why is this book for you?

As an information technology expert with nearly 50 years of experience of driving digital transformation and now AI initiatives, I have witnessed firsthand, both the opportunities and risks involved. Through this book, I aim to provide a practical resource for policymakers, business leaders and technologists navigating the complex landscape of AI in Africa. The book will present real-world case studies, assess some policy frameworks, and offer recommendations for maximizing AI benefits while mitigating risks.

We're going to look at the landscape of AI across Africa. We'll talk about some of the existing projects and available infrastructure. We will look at some examples of what AI has already done and what it is doing in Africa. We will examine some of the challenges, particularly in Government policy, providing the right enabling environment and infrastructure. We will see the importance of the engagement of our regional, and continental organisations, as well as inputs from the private sector.

The book aims to provide African policymakers, business leaders, and technologists a practical guide to navigating the complex AI landscape in Africa.

It will present real-world case studies, assess policy frameworks, and offer recommendations on maximizing AI benefits while mitigating risks.

The book will examine progress, promises, and perils of AI in Africa, including existing projects, infrastructure, and challenges around policy and enabling environments. It emphasizes the need for engagement from regional/continental organizations and private sector.

The book aims to inform strategies, policies, and decisions to help Africa lead in developing AI technologies that solve local challenges while reflecting African values globally.

The book also provides aspiring leaders in the field of AI in Africa with examples of what is already working and why. This book shows how you can build a career in AI and if you are already in a leadership position, how you might take better decisions. For those about to enter this field or planning to retrain, we show where some of the best schools can be found.

We will end with some conclusions and a look to the future. African stakeholders need to lead in shaping, not just AI adoption locally but to also contribute to developing ethical, responsible AI norms around the World. Increasing investments including data centres and telecommunications infrastructure across Africa reflect both public and private sector efforts underway to expand data capabilities. However, successful applications of AI remain limited in scale and scope.

This short book investigates the progress, promise and perils of AI in Africa today. There are positive examples from all over the continent, as well as indications of where problems may be found. We cannot include everything, but collectively, these chapters provide a fairly comprehensive perspective on the present landscape, future possibilities and potential pitfalls for AI in Africa.

The time for action is now if Africa is to lead in developing AI technologies that solve local challenges while reflecting her values on the global stage. This book aims to inform strategies, policies, decisions and actions that help realize this future.

We could not cover everything and everyone in AI on that vast continent. Apologies if we have not mentioned your favourite AI project. Africa is a big place and needs even more investment in AI.

But as you go through this book, please see what you can learn, from the start up, or mature AI project or the policies of government. What does the AI journey of those in this book tell you about your own roadmap? Who have they partnered with? Who has funded them? Each fact, figure and story is there to help you and me think about our own possibilities.

CHAPTER 1- WHAT IS AI?

The Core Idea in Chapter 1:
Artificial Intelligence (AI) offers major opportunities for Africa's development but strategic investment in skills, data, and responsible governance is urgently required.

AI Explained

AI is the effect of programming computers so that they imitate or appear, to think and learn like humans. The reason for developing Artificial Intelligence (AI) in different areas of life, is simply the need to solve problems, making our tasks easier.

All forms of AI use algorithms which are increasingly complex computer programs or sets of instructions. These Algorithms are descended from the old-fashioned Algebraic equations you remember from school. They were then further developed with the addition of statistical methods, logical reasoning, classification, clustering and predicting outcomes. Their development continues... New versions of AI based apps are being announced almost every day.

Researchers and developers in the field of Artificial intelligence have a great suite of specialized programming languages[5]. One of the most widely used coding languages across many areas of AI is Python. Knowledge of coding and Data Management is a good foundation for working with AI, but not every role needs that level of technical expertise. However, there are web pages online with information on the use of Python coding in AI and Machine Learning[6]. There is more

---

5

https://en.wikipedia.org/wiki/List_of_programming_languages_for_artificial_intelligence#cite_note-6

[6] https://wiki.python.org/moin/PythonForArtificialIntelligence

on your options in education for AI (including Python and other coding) and how to start a business in AI, later in this book.

But put simply, algorithms help machines look for, or follow patterns in sets of data. All areas of AI use algorithms in their different approaches especially in training machines, to automate tasks currently still performed by people. This is the reason that the world of work (and even your home and leisure) is already changing rapidly and will soon change even more. Africa should get ready.

Currently, there are two main types of AI – Narrow (Weak) AI and General (Strong) AI. Weak AI is represented by Siri or Alexa and many other gadgets we commonly use. Weak AI usually has a narrow or simple set of tasks that it can carry out. General or Strong AI is expected to be "more human". It still gradually developing, and we do not have good examples yet.

Big Data
Modern technology means that we now live in a world full of all kinds of Big Data that is growing more, and faster every day. Big Data is simply too large for the old data-processing methods, but developments in computing now make handling these levels of data possible. Africa with her large land area and huge population, needs to harness AI, Big Data, and Machine Learning as a priority.

We need AI to extract useful information from Big Data. Then AI also needs Big Data to learn how to find those patterns in large sets of data. They need each other. What AI learns can then also become forms of Machine Learning (ML), speech recognition/synthesis and understanding problems and processes etc. We then say that the computer has gained a form of "Artificial Intelligence" or "AI."

Google Search and other sophisticated web search engines are increasingly using AI. We are moving from search engines (Google and Bing) to reasoning engines (Chat-GPT, Bing Chat etc.). YouTube, Amazon, Netflix and other suggestion systems help you choose your options. Siri and Alexa - can understand human speech to take instructions. The latest version of Alexa has been upgraded with AI.

AI is successfully helping radiologists scan through huge numbers of x-rays and other results, picking out the ones that might need further investigation. Self-driving cars are achieving various new levels of autonomy. ChatGPT and AI art - generative or creative tools are flooding the market. We all also know that computers have long been programmed for games like chess. Historically, AI has included several areas of AI development, some of which overlap with each other.

Just a very brief comment about whether AI is new or not. Decision making systems have been around for a long time. The huge difference we are seeing with AI now is the mass of data that it can deal with. This is what is bringing out a new generation of "intelligent machines".

Increasingly whatever computing or machine process you use, from money transfers to GPS, it is likely that AI will be doing part of it. Many of these systems had already been available, but AI has injected a massive new level of capacity.

Artificial Intelligence is an area of computing that is growing and changing very fast. AI is likely to become even more sophisticated in the coming years. There are handy outlines of key skills needed for the AI/ Machine Learning Entrepreneur, a tools for measuring the readiness of your ML Project, and a list of further reading at the end of this book.

Meet the Continent of Africa

From: www.freeworldmaps.net/printable/africa/africa_countries.jpg

Coming in behind Asia (which includes giants China and India), Africa is the second-largest continent in both size and population. Including the African islands, it covers roughly 30.3 million km² (11.7 million square miles) or 6% of the Earth's surface and 20% of its land[7].

---

[7] Sayre, April Pulley (1999), Africa, Twenty-First Century Books

Despite being classed as a "poor" continent, Africa has historically provided great wealth to most of the world's empires and businesses, including minerals, agriculture, and her people.

On Sunday, September 24, 2023, and based on United Nations estimates, the population of Africa was one billion four hundred sixty-eight million four hundred twenty-five thousand six hundred twelve people (1,468,425,612), or 17.89% of the total world population[8].

Big Data needs such large populations, but still requires well-structured methods of gathering processing and harvesting information from that data to prevent "big bias"[9]. Later in the book we will see how this size and population should make Africa the next global AI superpower after China and the USA. We will ask what needs to be done to achieve that and what the roadmap might look like.

According to Saifaddin Galal of Statista, the largest cities in Africa in 2023, by number of inhabitants are:

- Lagos, Nigeria, 9m
- Kinshasa, Democratic Republic of the Congo, 7.8m
- Cairo, Egypt, 7.7

See more in the graphic below.

---

[8] https://www.worldometers.info/world-population/africa-population/ (Accessed 24/09/2023)

[9] Kaplan RM, Chambers DA, Glasgow RE. Big data and large sample size: a cautionary note on the potential for bias. Clin Transl Sci. 2014 Aug;7(4):342-6. doi: 10.1111/cts.12178. Epub 2014 Jul 15. PMID: 25043853; PMCID: PMC5439816. (Accessed 24/09/2023)

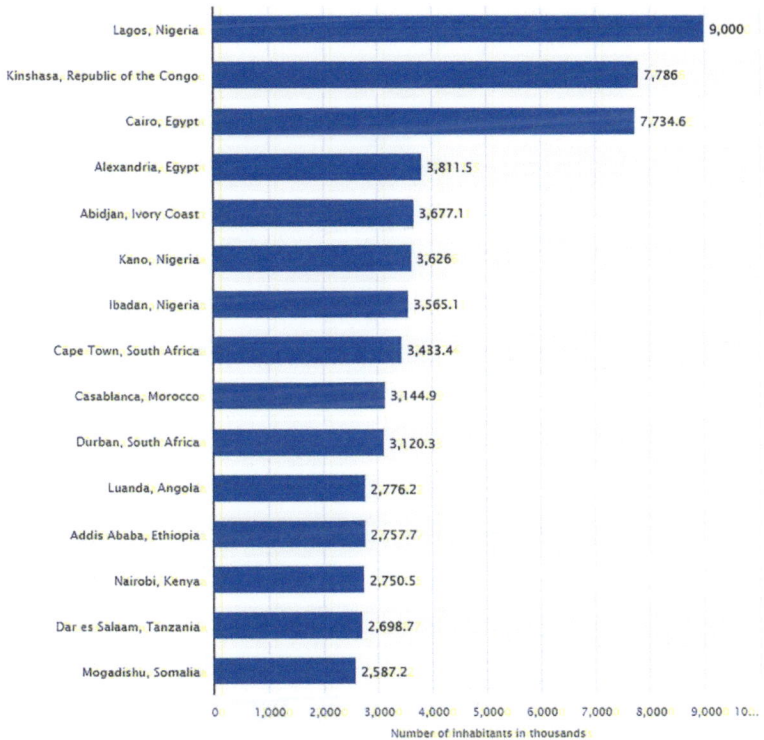

Largest cities in Africa 2023 by number of inhabitants (statista.com)

Since 2000, urbanization across Africa has increased to 35 percent and the urbanization rate in Africa was anticipated to reach 44 percent in 2021. However, this trend varies across the continent[10]. By the end of this century Africa is predicted to have about 40 percent of the global population with nearly 4.3 billion people[11]. This creates a real need for

---

[10] https://www.statista.com/statistics/1218259/largest-cities-in-africa/ (Accessed 23/09/2023)
[11] https://population.un.org/wpp/ (Accessed 23/09/2023)

harnessing all available technology to the best benefit of that huge population, including the best aspects of Artificial Intelligence.

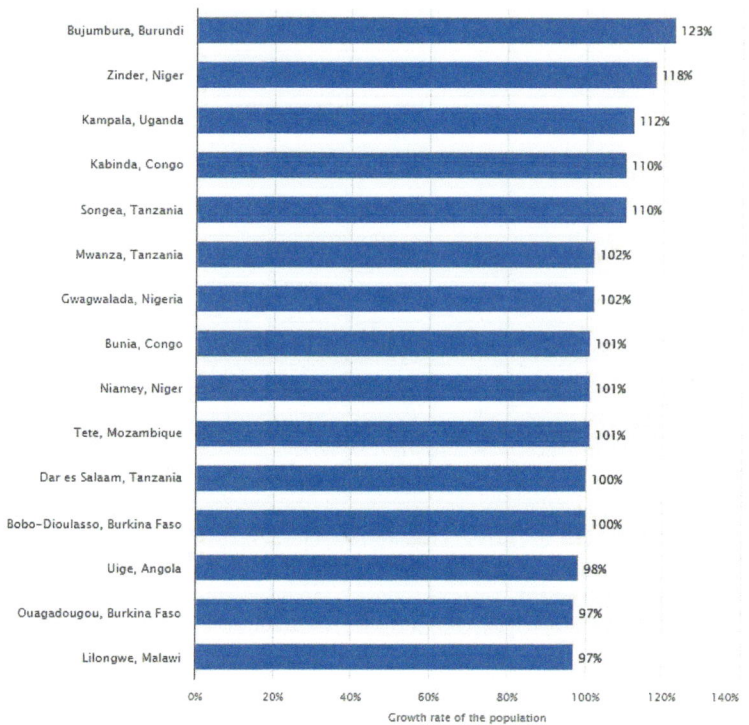

Fastest growing cities in Africa between 2020 and 2035 (statista.com)

The median age in Africa is 18.8 years. Africa has remained the youngest continent by average age since 1955 and is set to remain so. For Africa to reap the benefits of AI, the technology gap between

African youth and those in other more developed countries needs to be closed[12]. Could Africa be missing her biggest opportunity?

Many organisations are already gathering and processing Africa's data, with some benefits to the continent. But if data is the new and most precious resource, who is regulating the mining of Africa's data? Later in this book, we will assess the "scramble for African data".

---

[12] Harry, Njideka U. (11 September 2013). "African Youth, Innovation and the Changing Society" https://www.huffpost.com/entry/african-youth-innovation_b_3904408 (Accessed 24/09/2023)

AI- An African Opportunity

Artificial Intelligence (AI) is poised to transform economic development across Africa over the coming decade. From healthcare to agriculture, finance to education, AI solutions are already demonstrating enormous potential to help tackle the major challenges currently constraining Africa's growth.

However, along with the hype, Africa must prepare for her own version of the major changes in the job market and daily life that are already sweeping across the world.

To give just one example here, David Vernon, in "Robotics and Artificial Intelligence in Africa" (2019) and following J. Manyika et al., (2017), predicts a workforce displacement by automation for Kenya of 5.5%, Nigeria 8.5%, and South Africa 12.5%[13].

Serious ethical and governance concerns remain around creating an inclusive, human-centric AI ecosystem that is based firmly in the context of African societies and cultures.

Susan Brokensha, Eduan Kotzé and Burgert A. Senekal have produced an excellent in-depth book titled, "AI in and for Africa: A Humanistic Perspective". Brokensha and her co-authors are all of the University of the Free State, South Africa and their book is suitably heavyweight academic reading[14].

---

[13] Vernon, David, Robotics and Artificial Intelligence in Africa, IEEE Robotics & Automation Magazine, Pp 131-145, December 2019

[14] Brokensha S., (Auth), Kotzé E., (Auth), & Senekal B. A., (Auth), AI in and for Africa: A Humanistic Perspective, Publisher: Chapman and Hall/CRC, 1st Ed., 2023. https://www.routledge.com/AI-in-and-for-Africa-A-Humanistic-Perspective/Brokensha-Kotze-Senekal/p/book/9781032231761

They echo the calls of several other authors that AI in Africa must be contextualised and made indigenous to Africa. This they say will help evole a unique socio-technical apporach to AI in Africa. They believe that a more humanistic view of AI allows for cultural, ethical, and ecological dimensions to become central to AI right from the to the development stages.

The authors arc also in agreement with me that if not handled well. the advent of AI in Africa might be skewed to benefit a few, while marginalising many. They agree with my own view that AI and indeed the Internet is inherently and covertly "White". Like many other writers before them, Brokensha Et al., see the introduction of modernising technologies into Africa as too often entrenching colonial legacies of inequalities.

Strategic interventions are needed to build world-class technical talent and data assets that fuel homegrown innovation before Africa falls further behind as the pace of technology change grows faster by the day.

CHAPTER 2 – AI INVESTMENT IN AFRICA

The Core Ideas in Chapter 2

AI investment in Africa from global tech companies and multinational agencies, is surging but needs to increase, and responsible governance frameworks are needed to create truly inclusive AI ecosystems.

Across Africa, Artificial Intelligence (AI) projects and applications are already emerging to transform major sectors. These innovations are offering improved efficiency, insights, and process automation. In healthcare, machine learning is being applied to predict and map disease outbreaks.

In education, AI-driven adaptive learning platforms have begun to provide customized lessons helping students gain literacy and numeracy. However, such benefits are still often for the very few who can afford them. Financial institutions are also using AI algorithms to provide credit scoring and detect fraud, expanding access to capital.

The economic impacts of these innovations promise to be profound. Startups applying AI solutions have attracted major investments from global tech giants like Google and Huawei looking to tap the booming young population. The World Economic Forum projects AI could create 58 million new jobs in Africa by 2022 as it reshapes industries.[15]

Africa's AI Investors

There are already many examples of AI across Africa, in:

- Fintech

---

[15] https://www.weforum.org/press/2018/09/machines-will-do-more-tasks-than-humans-by-2025-but-robot-revolution-will-still-create-58-million-net-new-jobs-in-next-five-years/ (Accessed 26/09/2023)

- Healthcare
- Agriculture
- Retail/E-commerce
- Robotics, and Autonomous vehicles.
- Others

Let's meet some of these AI investors and their investments.

Strive Masiyiwa

In September 2023 the Founder and Executive Chairman of African telecoms industry pioneers Econet, Strive Masiyiwa sat down for a chat about AI in Africa, with James Manyika, Google's SVP of Research, Technology and Society, in the "Dialogues on Technology and Society series Episode 2". [16]

Strive Masiyiwa. Picture Credit: Kathi Walther Bouma[17]

---

[16] Strive Masiyiwa and James Manyika, Dialogues on Technology and Society, Ep 2: AI and Africa, YouTube Sep 21, 2023 (Accessed 25/09/2023)
[17] CC BY-SA 2.0 <https://creativecommons.org/licenses/by-sa/2.0>, via Wikimedia Commons

Seasoned African entrepreneur and philanthropist Strive Masiyiwa shared his thoughts on wide ranging issues within and around the development of Artificial Intelligence across Africa.

Strive Masiyiwa sees the mobile phone revolution as having had an extraordinary impact on Africa, increasing phone access from 0.7% to over 80% of the population. He believes it has led to major GDP growth in many parts of Africa. He is therefore also excited about the possibilities of AI for Africa, often enabled by handsets, like providing equal access to education through AI teachers for example. He believes AI can help solve major challenges like education inequality.

However, Strive Masiyiwa has concerns about the risk and challenges of AI. He worries AI could amplify inequality if Africa does not fully participate. He is also concerned about:

- Lack of investment
- Math and science education gaps
- Potential resentment if benefits of AI flow elsewhere.

Strive Masiyiwa wants Africans to be full economic participants in AI, not just users of AI platforms built by others. He wants to see more African entrepreneurs building AI enterprises and the wealth that can come from that. He explained that he has invested heavily in digital infrastructure like fibre networks, data centres and renewable energy, helping to ensure that Africa has the foundations for AI.

However, he is not yet comfortable supporting open-source AI until more frameworks are in place for responsible management and control and also believes that we need to be both bold and responsible with collective responsibility from entrepreneurs, governments and society to get African AI right.

Strive Masiyiwa sees major potential for AI as a tool to bring equality to education and reach underserved African children. He pointed to the growing excitement and entrepreneurial energy around AI already emerging across Africa with examples like AI research centres in Ghana and Rwanda. But he says that more participation is still needed.

The African Development Bank

The past two years have seen several investments in AI across Africa, with significant support from the African Development Bank (AfDB). In October of 2022 Rwanda started working with Viebeg's data-driven logistics platform[18]. The project was funded by the Rwanda Innovation Fund and supported with $30 million (equity) by the AfDB.

---

[18] https://www.afdb.org/en/success-stories/how-rwanda-using-artificial-intelligence-improve-healthcare-55309 (Accessed 23/09/23)

The Viebeg AI system, which is expanding in Eastern Africa, is designed to manage supply chains, and connect healthcare directly with manufacturers, thereby reducing costs and stabilizing supply.

The United Nations Conference on Trade and Development Economic Development in Africa Report 2023 is titled "The Potential of Africa to Capture Technology-Intensive Global Supply Chains". The report highlights the increasing investment in digital technology and skills across the continent which makes vertical or horizontal integration af African businesses into global supply chains possible[19].

In January 2023 the AfDB approved another $20 million equity investment of in the Africa50 Infrastructure Acceleration Fund [20]. This is a pan-African private equity infrastructure fund that is mobilizing up to $500 million for digital and other infrastructure investment. Many of these projects will be a Public Private-Partnership (PPP).

In September 2023, during the Global Africa Business Initiative at the United Nations General Assembly in New York, the AfDB and Google signed a Letter of Intent (LOI) on digital transformation in Africa[21]. The LOI focussed on emerging technologies, upgrades to infrastructure, and talent and skills development across Africa.

---

[19] https://unctad.org/node/42171 accessed 25/09/2023
[20] https://www.afdb.org/en/news-and-events/press-releases/african-development-bank-approves-20-million-investment-private-equity-fund-targeting-infrastructure-sector-africa-61683 (Accessed 230923)

[21] https://www.afdb.org/en/news-and-events/press-releases/african-development-bank-and-google-collaborate-digital-transformation-africa-64488 (Accessed 230923)

According to their website, the AfDB has invested $1.9 billion over the past decade in projects for "...the development of broadband infrastructure, conducive policy and regulatory environments, digital skills, and innovative technology startups". Google also has many other investments in Africa's digital development, several of which we shall see shortly.

African Development Bank also provides over $1 million for AI-based multi-lingual banking systems in Africa[22]. These will process customer complaints for the Competition and Consumer Protection Commission of Zambia and the national banks of Ghana and Rwanda.

The grant is from the Africa Digital Financial Inclusion Facility (ADFI), a special fund to step up digital financial inclusion across the African continent. Housed and managed by the African Development Bank. ADFI is a multi-donor trust fund established as a blended finance vehicle[23], with the following partners:

- African Development Bank (AfDB)
- Agence française de développement (AFD)
- Bill and Melinda Gates Foundation (BMGF)
- French Treasury at the Ministry of Economics, Finance and Industrial and Digital Sovereignty
- Ministry of Finance, India
- Ministry of Finance, the Government of Luxembourg

---

[22] https://www.afdb.org/en/news-and-events/press-releases/african-development-bank-provides-1-million-ai-based-national-customer-management-systems-ghana-rwanda-and-zambia-42602 (Accessed 230923)

[23] https://www.adfi.org/about-us/overview (Accessed 25/09/2023)

- Women Entrepreneurs Finance Initiative (We-Fi).

Developers of the AI system are Sinitic Africa the subsidiary of Canadian financial technology firm Sinitic Inc., and BFA Global, an impact innovation consultancy. BFA specialize in human-centred design and digital financial services (DFS) regulation of financial services to unbanked people globally.

UNICEF, Google, Microsoft & more...

An innovative UNICEF project in Cote d'Ivoire captioned "Bringing HIV Prevention into the Twenty-First Century - U-Test" utilizes geo-mapping, social media, and AI[24]. This social protection project is set up as an HIV prevention outreach to young people.

Although the project seems successful, there could be questions about the data profiling of these young people in spite of attempts at data anonymisation as we shall see later in this book. UNICEF has many other projects around Africa and across the world.

Google has a long track record of digital investment in Africa, notably in the major Seacom submarine telecommunications cable which has

---

[24]   https://www.unicef.org/innovation/stories/bringing-hiv-prevention-twenty-first-century (Accessed 24/09/2023)

connected Djibouti, France, India, Kenya, Mozambique, South Africa and Tanzania since 2005. Google is another partner of the AfDB[25] and has pioneered and partnered several other innovation, infrastructure, skills building and governance initiatives, that have been significant in the gradual digital transformation of Africa.

Google's Senior Vice President of Research, Technology & Society, Dr. James Manyika, has predicted that Ai will be the game changer for Africa.

In June 2019, the Voice Newspaper was one of many media to report the opening of Google's Artificial Intelligence (AI) research centre in Accra Ghana, the first in Africa[26]. According to the report, head of Google AI, Moustapha Cisse said the centre would work across the continent, specializing in machine learning research. The centre also aims to expand information access in African languages.

Since opening an AI research center in Ghana in 2018, Google has made progress through collaborations with universities, governments, and local communities. Nurturing local AI talent in Africa is critical. Google partners with African universities to run AI residency programmes.

Another example of a Google project is using satellite imagery and AI to map Africa's built environment (Open Buildings initiative). One

---

[25] https://www.afdb.org/en/news-and-events/press-releases/african-development-bank-and-google-collaborate-digital-transformation-africa-64488 (Accessed 230923)

[26] https://www.voice-online.co.uk/news/tech/2019/06/12/google-opens-first-african-ai-lab-in-ghana/

open-access dataset project based on satellite imagery is pinpointing the locations and geometry of buildings across Africa, Open Buildings allows the team to study power distribution gaps and potential solutions, in detail. This data helps with urban planning, energy needs, and crisis response.

Conversational AI has potential for education. Google has also built an interview prep tool to help African job seekers. Climate change and sustainability are a big focus of their work. AI helps predict floods and locust outbreaks that threaten food security.

Google aims to apply AI solutions to agriculture, education, and healthcare among others. Google has other AI centres across the world, including Montreal, New York, Paris, Tokyo, Zurich, and Tel Aviv/Haifa.

Solutions built in Africa can scale globally. Machine translation of African languages expands information access. There are still challenges but huge opportunities exist in Africa given the young population. Continued innovation and collaboration with partners can make a profound impact.

One question we will still need to always ask, is what are the data protection frameworks for the data collected in this and other projects?

In January 2019, building on an MOU from June 2018, First Bank of Nigeria Limited, and Microsoft 4Afrika launched a new partnership[27]. Together they offered easy access to Microsoft products and skills building to accelerate the digital transformation of Nigerian SMEs who are customers of First Bank. Microsoft 4Afrika has similar partnerships with banks, telecoms, and other players across Africa.

Also in 2019, quantum computing pioneers IBM entered a partnership with

---

[27] https://www.firstbanknigeria.com/microsoft-4afrika-partners-firstbank-to-support-smes/

Wits University in Johannesburg, South Africa[28].

This expanded the IBM Quantum Computing program to Africa. The partnership plans to increase education and research opportunities with the more than 15 members of the African Research Universities Alliance (ARUA).

The Chinese multinational Huawei plans to train 10,000 Africans in AI at the Huawei ICT Academies across the continent over 5 years. Graduates will receive Huawei certification[29].

However, some of the other biggest investors in artificial intelligence (AI) in Africa we should notice are:

- Intel - Has invested over $100 million in AI in Africa since 2015. Runs the Intel AI for Youth program teaching AI skills to young people.
- Facebook - Invested in AI research collaborations with universities like WITS in South Africa and runs AI training programs across Africa.
- Orange - Telecom company that launched a $56 million Orange Digital Ventures Africa fund to invest in AI startups.
- JUMO - Africa-based fintech startup using AI for financial services that has raised over $200 million.

---

[28] https://newsroom.ibm.com/2019-06-12-IBM-Expands-its-Quantum-Computing-Program-to-Africa-with-Wits-University

[29] https://www.huawei.com/za/news/za/2019/huawei-argets-10000-south-african-students-for-ict-competition-2019-2020

- Andela - African startup applying AI for talent recruitment and skills training that has raised over $180 million.
- Africar Group - African conglomerate that launched a $100 million venture fund to invest in AI companies.

Growth in African AI Investment

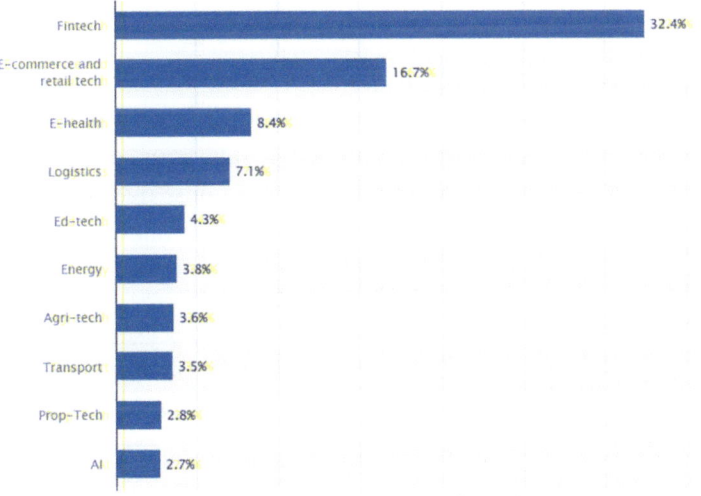

Distribution of funded tech startups in Africa (From statista.com)

The 2021 Africa Tech Venture Capital Report by Partech shows that this was the 6th straight year of growth in African technology Venture Capital with an influx of new investors[30]. $5.2 billion was raised by 640 African tech startups across 681 equity funding rounds. This was a

---

[30] https://partechpartners.com/africa-reports/2021-africa-tech-venture-capital-report

264% increase in funding and a 92% increase in deal volume compared to 2020. There were 21 Megadeals deals (above $50M) in 2021. Deals below $200K were excluded from the study.

The highest number of deals went to Nigeria, with most of the investment going to fintech. Partech Africa, Anza and other funds have reportedly invested in AI startups DataProphet, TalentQL, and InstaDeep. There are many others, probably too many for us to try to cover here. However, overall funding for Ai in Africa is still low compared to other technologies.

mPharma
mPharma is a Ghanaian health tech startup rolling out virtual medical centers across 7 African countries[31]. Their goal is to provide quality primary care through virtual consultations and remote medical examinations at their pharmacy locations. mPharma already offers 10,000 in-person doctor consultations at pharmacies and wants to transition these to electronic visits.

The virtual centers will allow patients to get examined remotely using digital medical devices during video consultations. mPharma aims to drive the growth of telemedicine in Africa as mobile connectivity expands. The startup has raised over $50 million in funding, including $17 million in 2019.

The startup plans to create more jobs to support the expansion into virtual care. Telemedicine has big potential in Africa given low doctor-to-patient ratios and deepening mobile connectivity. mPharma hopes to partner with other telemedicine providers to utilize their infrastructure and examination capabilities.

---

[31] https://mpharma.com/

Other Medical Uses of AI in Africa

South Africa - The Council for Scientific and Industrial Research (CSIR) a leading scientific and technology research organisation has launched Project AID[32]. The new techniques utilize Ai capabilities in detecting cases of Tuberculosis (TB). An Act of Parliament established the CSIR and it remains under the Minister of Higher Education, Science and Innovation. Babyl Rwanda is also working for better neonatal care using AI and robotics[33].

Other Notable Investments

Others we could include are Intsar and Karin in Egypt, Recommendio in South Africa, and Zipline in Ghana and Rwanda, and Delivery Robots in Uganda. We will meet some along the way and in the next chapter, we will hear about the Pasteur Institute, and their partners in Senegal.

AI unquestionably holds enormous potential as a catalyst for economic and social progress in Africa. But as nations look to harness its power, they must also safeguard core values around equity, ethics and human development so that AI works for all segments of Africa's diverse populations.

---

[32] https://www.csir.co.za/csir-driving-smart-innovations-boost-sas-healthcare-system-rural-areas (Accessed 27/09/2023)
[33] https://www.babyl.rw/ (Accessed 27/09/2023)

CHAPTER 3- AI CASE STUDIES IN KEY SECTORS

The Core Ideas in Chapter 3:

Case studies have proven the success of AI applied for social good in African healthcare. Investment in AI skills can help create new technical jobs across the workforce.

Healthcare- Predicting Disease Outbreaks

In Senegal, the Pasteur Institute collaborated with local and international partners to develop a machine learning model for predicting meningitis outbreaks. Meningitis causes deadly epidemics in Sub-Saharan Africa's meningitis belt.

Young African cattle farmer

First proposed in 1963 by Léon Lapeyssonnie of the World Health Organization (WHO). The African meningitis belt is now recognised as stretching from Senegal to Ethiopia[34]. Until this Pasteur Institute project, early warning systems were inadequate to target vaccination efforts.

The Pasteur Institute has been In Africa for a long time. You may have your own opinion about them and even more generally about the French Government and other organisations from former colonial powers in Africa. This book is not a defence of them or other such organisations, but certainly Pasteur Institute is one of those who have made huge contributions to healthcare. They have been in Africa since the late 1800s and they're still around. However, in this season of AI development across Africa, we may need careful data governance.

By analysing 50 years of data from 10 sub-Saharan countries on disease incidence, climate and geography, the Pasteur Institute model achieved 85% accuracy in forecasting outbreak location and timing 2-5 months out. This has enabled better preparedness saving thousands of lives. The success demonstrates the power of AI for social good when public and private partners share data and expertise.

Using such historical data needs careful handling. Many of the records will be paper based, such as notes taken by doctors or other clinicians, or prescriptions. It might be of varying quality and might even contain

---

[34] Lapeyssonnie, L. (November 1968). "[Comparative epidemiologic study of meningococcic cerebrospinal meningitis in temperate regions and in the meningitis belt in Africa. Attempt at synthesis]". Médecine Tropicale: Revue du Corps de Santé Colonial. 28 (6): 709–720. ISSN 0025-682X. PMID 5739513.

inherent bias from different historical and social times, such as the colonial era.

There will need to be a careful process of legacy paper record digitization before any other data processing and eventual integration can begin. The Pasteur Institute successfully integrated the historic data with the other data sets, and then they started to train the AI system to become a predictive model. They have successfully used Artificial Intelligence, including Google AI to predict the spread of meningitis.

To achieve this, the Pasteur Institute have had help. The meningitis prediction project was conducted with other international collaboration. The global diagnostic biotech company bioMérieux Foundation and Dataiku, (the AI and ML unicorn) were also partners.

St Louis Market Senegal

This and other collaborations have helped build local capacity so that medical specialists and others within that area have been able to grow from this experience.

The Pasteur Institute were able to correlate disease incidents of the cerebral spinal meningitis, outbreaks with geography, climate change, and many other factors which gave the governments and the medical specialists within a very good picture of what was happening and what that was related to.

AI is able to crunch masses of data in ways that it'll be very difficult even if you keep flying around with a helicopter. You simply can't break data down into information like that very easily. The output was an updateable predictive model that changes the outcomes in the lives of the real stakeholders, the citizens on the ground.

Updateable predictive models can be retrained to accommodate changes in situations. This project also achieved a forward accuracy of five months in estimating both probability and geographic spread. That means that the system alerts the health teams that in about five months there will be a serious Meningitis outbreak and they can activate a planned vaccination procedure and other protocols.

This is another good example of some of the ways in which healthcare and particularly disease prevention could be enhanced across the African continent.

AI For Job Creation

accenture

Despite many public doubts, South Africa's government is investing in AI skills development. A report published in 2021 by Accenture in collaboration with the Gordon Institute of Business Science (GIBS) at the University of Pretoria says public readiness for AI in the Republic of South Africa is lagging, with concerns around data quality, privacy, workforce skills, and potential job losses. There is also a lack of competencies to fully integrate AI into the economy[35].

The authors of the report are Willie Schoeman, Rory Moore, Yusof Seedat, and Dr Jeff Yu-Jen Chen. Other contributors are Ntombi Mhangwani, Paul Barbagallo, and Madhu Vazirani. The report however suggested that:

- AI has potential to add 1 percentage point to South Africa's annual economic growth by 2035.
- South Africa needs to create a vibrant AI ecosystem across universities, startups, large companies, policymakers and partnerships.
- Companies should focus on AI invention and collaboration to drive growth.
- Responsible AI practices are needed around governance, ethics, codes of conduct.

---

[35] https://repository.up.ac.za/handle/2263/82719

- Policymakers and business leaders need strategies for workforce readiness, infrastructure, fostering collaboration.

Although South Africa has a good record in encouraging AI start-ups and attracting investment into the sector, the government seems to be laying a greater emphasis on capacity building, much of that through education. Here are some examples of how educating South Africans in AI can help:

Multi-Dimensional Interventions

The Department of Science and Innovation's new institute for AI research will build capabilities and skills for students and professionals to pursue careers in AI. It aims to foster collaboration between universities, companies, and government on AI R&D.

The National Research Foundation provides funding for Masters and PhD students at universities across South Africa to study AI, machine learning and data science. This grows a talent pipeline for technical roles.

The Expanded Public Works Programme is training unemployed youth in data science and machine learning skills so they can find jobs as AI technicians, data analysts, machine learning engineers etc.

The Innovation Hub in Pretoria runs an accelerator program for early-stage AI startups, providing training in business and technical skills. This boosts entrepreneurship and jobs in the local AI industry.

Provincial academies of science sponsor challenges, hackathons and workshops in AI and robotics to spark interest among high school learners. This exposes youth to pursue further education and careers in AI.

Major universities like Wits, Pretoria, Johannesburg have new dedicated research centres, undergrad and postgrad degree programs in AI, computer science and data science.

Partnerships between universities like Wits and tech companies like Google provide funding, mentorship and internships in AI to help students gain practical experience.

Tax Incentives for R&D: The South African government also offers a deduction for companies doing research and development in 4IR technologies like AI and machine learning. This encourages private sector investment in developing innovations and capabilities.

New Vocational Programs: The Department of Higher Education and Training is introducing new vocational programs focused on data science, machine learning and AI at technical and vocational education and training (TVET) colleges across South Africa. These 1-2 year diploma programs will provide practical training to youth to gain in-demand digital skills for employment as data analysts, AI technicians, etc. Several TVET colleges like Eastcape Midlands, Ekurhuleni West, and Sedibeng are piloting these programs starting in 2023.

So through R&D tax incentives and new vocational programs, South Africa aims to spur development of AI innovations by companies and equip youth with practical data science and AI skills for the job market. The vocational programs in particular will provide an alternative to traditional university education to make emerging digital skills more accessible.

The African Institute for Mathematical Sciences (AIMS) runs data science training programs. Established in September 2003, (AIMS) is a tertiary education and research institute in Muizenberg, South Africa.

It has an associated network of linked institutes in Senegal, Ghana, Cameroon and Rwanda

Government funding aims to spur innovation while also expanding job opportunities. The goal is to build both high-level skills for research as well as more broad-based capabilities to apply AI tools and technologies.

So, through various initiatives in academia, government and industry, educating South Africans in AI skills at all levels will grow the country's talent pool to fill critical jobs in the AI economy. We can also say that many of those being trained in AI today will go on to become providers of jobs for others in the future.

CHAPTER 4- DATA PROTECTION AND AI GOVERNANCE

The Core Idea in Chapter 4:
Responsible and contextualised AI governance must be built into the future of AI in Africa.

Human Development and Political Instability

The track record of AI development in other parts of the world, shows Africa that good data protection and data governance frameworks are essential.

Africa is one of the more fragile political parts of the world. Research has shown what even the most casual observer might conclude that political instability and low human development are very often closely related. Can AI bring positive change without negative risk?

The United Nations Development Programme is a UN agency that helps countries reduce poverty and achieve economic growth and human development.

The UNDP focuses on building local capacity for long-term self-sufficiency. The Human Development Index (HDI) measures achievement in health, education, and living standards. Health is measured by life expectancy, education by years of schooling, and living standards by Gross National Income (GNI) per capita.

The HDI aggregates these into a composite index using a geometric mean. The HDI can be used to compare human development outcomes between countries with similar income levels and inform policy debates.

The map below shows the HDI of African countries. The so-called Coup Belt of mainly former French colonies is highlighted in red. Following the 2023 coup, Gabon can be included. Apart from Gabon, the HDI of all the Coup Belt countries is below 0.500.

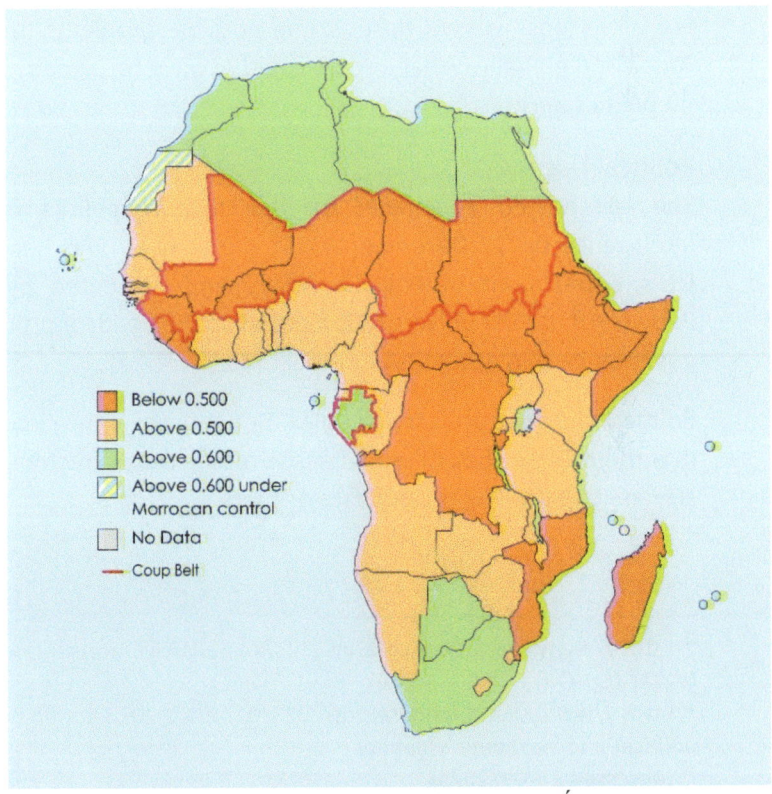

HDI of every African country, Gabriel A. Álvarez N. from Wikipedia (Creative Commons) 2 September 2023

Aleix Montana is a Research Associate at Verisk Maplecroft. Drawing analytical conclusions from quantitative data in the Verisk Maplecroft Civil Unrest Index, he observes that between 2022-Q2 and 2023-Q2 the region has seen the largest increase in risk since 2017[36].

Dr Alex Vines OBE, Research Director; Director, Africa Programme at Chatham House believes that Africa's high inflation and interest rates, exacerbated by domestic and foreign crises, will lead to ongoing political and economic instability in 2023[37].

For all its positives, AI poses some possible problems to politics in Africa and globally. What impact could AI have (or already be having) on politics in Africa?

Political Uses of AI

The UN hosted a seminar on the new geopolitics of artificial intelligence on 3rd December 2018. The seminar was hosted by the UN's Department of Political Affairs and presented by Eleonore Pauwels of the UN University. It focused on the need for prudent global governance of AI given its potential risks.[38]

Political elections are very complex. Depending on the circumstances, it is more likely that AI would assist rather than directly lead to the success of a particular candidate or party[39].

---

[36] https://www.maplecroft.com/insights/analysis/civil-unrest-in-africa-hits-6-year-high/
[37] https://www.chathamhouse.org/2023/01/africa-2023-continuing-political-and-economic-volatility
[38] (Accessed 25/09/2023)
[39] https://www.technologyreview.com/2023/07/28/1076756/six-ways-that-ai-could-change-politics/

The UN seminar outlined that AI has many positive use cases but could also be abused for harmful purposes like disinformation campaigns. Deepfakes and hacked social media could trigger dangerous political responses. Personalized disinformation could influence elections.

Flag of the United Nations

AI's data gathering abilities raise privacy concerns, potentially increasing inequality and enabling "cyber-colonization." In response, the UN University launched a platform for stakeholders to collaborate on AI governance and build trust.

The seminar discussed both the promise and perils of AI and the critical need for governance frameworks to ensure its ethical and safe adoption globally. The risks of AI require robust governance to safeguard stability nationally and internationally. Diverse and inclusive cooperation is key to implementing AI safely in Africa and across the globe.

Jeongki Lim Assistant Professor of Strategic Design and Management at Parsons School of Design summarised and commented on another UN Seminar titled "Generative Artificial Intelligence: What It Is, What It Is Not and What It Can Be for the United Nations" held in July 2023.

The seminar examined the role of current popular Generative AI like ChatGPT and how it creates human-like content like text, images, and videos. The rapid improvement in generative AI is staggering but the core technology has existed for decades. data analysis and AI are not new to the UN, but generative AI causes unease about adopting it more deeply.

Fears include job loss and distrust in letting an inanimate software make decisions affecting humans. A systematic approach is needed to increase confidence. The first step could be to use and test AI directly to understand its capabilities and limitations. UN personnel can evaluate if AI assistants could handle routine tasks and improve their work.

Benefits could include that AI image/video generation could aid multilingual communication and identifying fake news. The UN should help guide AI development to ensure that they are economically and culturally inclusive worldwide. The UN should work to help determine how to include AI benefits in a shared future for humanity it envisions.

While rapid generative AI progress causes concerns, the UN should actively test and evaluate it, shaping its development to align with UN principles of benefiting humanity inclusively.

The rise of AI is fuelling intense demand for data to train machine learning algorithms. This heightens the need for robust governance to ensure data privacy and prevent misuse.

Data Governance in Africa

Across Africa, a patchwork of regulations has emerged but lacks coherence. Ghana's Data Protection Act 2012 provides basic privacy safeguards but still lacks details for effective implementation. Nigeria, Kenya and South Africa have more comprehensive laws aligned to global standards. The pace of change is rapid. Just like others across the world, these and other Governments across Africa are struggling to catch up.

The Personal Data Protection Act came into effect in Tanzania in May 2023[40][41]. The Nigeria Data Protection Act 2023 was signed into law by President Bola Tinubu in June 2023[42]. After Mauritania became the 15th state to submit its ratification, the African Union Convention on Cyber Security and Personal Data Protection (The "Malabo Convention") was set to come into force in May 2023[43][44].

Data Protection Africa

A private company registered in the Republic of South Africa, the Applied Law & Technology (Pty) Ltd (ALT Advisory) runs Public Interest Advisory Services. Their wide-ranging provision includes an extensive

---

[40] https://dataprotection.africa/tanzania-personal-data-protection-act-comes-into-effect/ (Accessed 27/09/2023)
[41] https://abcattorneys.co.tz/wp-content/uploads/2023/05/Personal-Data-Protection-Act-of-Tanzania-Sheria-ya-Ulinzi-wa-Taarifa-Binafsi-Tanzania-2022-ABC-Attorneys.pdf
[42] https://dataprotection.africa/nigeria-president-bola-tinubu-signs-the-nigeria-data-protection-act-2023-into-law/ (Accessed 27/09/2023)
[43] https://dataprotection.africa/malabo-convention-set-to-enter-force/
[44] https://dataprotection.africa/wp-content/uploads/malabo_roadmap_Sept_2022.pdf

track record and expertise in emerging technology and innovation across Africa[45].

One of their major projects has been "Data Protection Africa[46]" which seeks to advance data privacy right on the continent through mapping the condition of data protection laws and policy. 55 members of the AU are covered in the project.

The project was launched in 2019 following research on a smaller group of African countries with the Open Government Partnership and has been supported by the:

- Strategic Advocacy Fund.
- Foundation Open Society Institute.
- Information Program of the Open Society Foundations.
- Luminate.
- Africa Mradi programme of the Mozilla Foundation.

The legal framework of each country's data protection was evaluated using 50 different factors in seven different areas.

1. Domestic Legal Commitments.
2. International Legal Commitments.
3. Scope of Application
4. Transparency.
5. Accountability.
6. Participation.
7. Automated decision making.

---

[45] https://altadvisory.africa/#discover (Accessed 27/09/2023)
[46] https://dataprotection.africa/ (Accessed 27/09/2023)

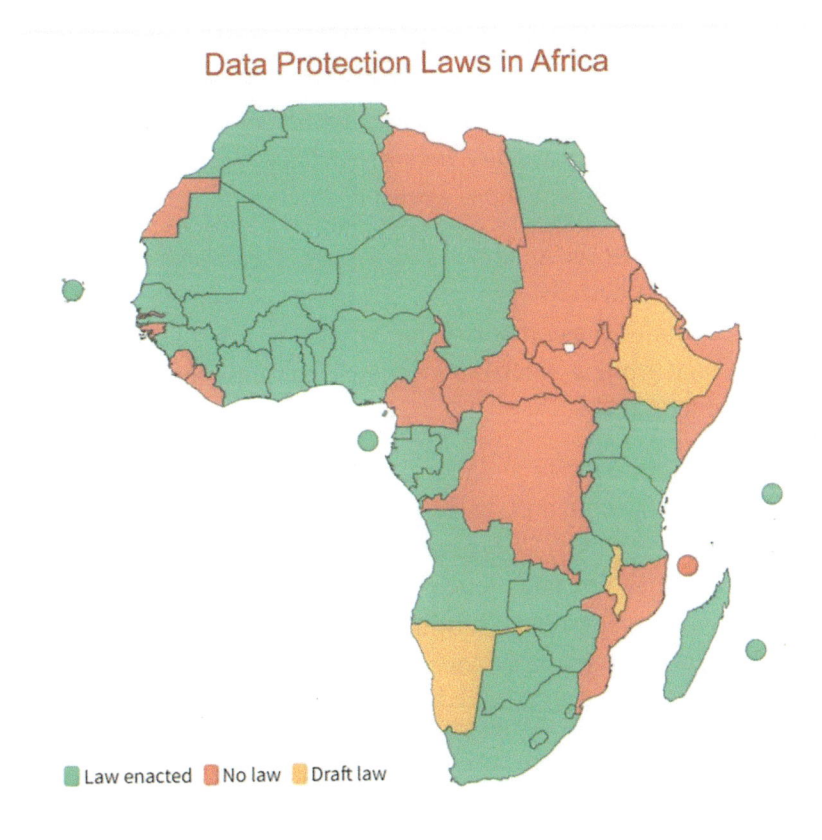

Data Protection Laws in Africa (From https://dataprotection.africa/)

AI Data Bias

There are some other key issues. We've talked about bias and discrimination. If we don't put transparency and fairness at the centre right from the beginning, we may end up with systems that are inherently unsuitable for Africa and for the values that are held across most of Africa.

Sunday Awoniyi of the Department of Religion and African Culture, Adekunle Ajasin University, Akungba-Akoko, Ondo State, Nigeria, describes these African social values as emphasising aspects like "...hospitality, chastity before marriage, truth, respect for old age, covenant keeping, hard work and good character"[47].

Bias can enter AI data in many ways. Those who write algorithms to build AI systems can also unwittingly introduce their own biases. Currently, those that develop Ai systems are mostly male, white, and come from a similar educational background of writing computer code.

Most of the data already being used to build AI systems to now had been collected and stored in Western societies based on responses of the groups most likely to answer survey questions. That means existing data may not have much input from different groups of minorities. Some developers are now building AI systems that can be prompted to reduce or remove a known bias[48].

AI systems that are already being brought into Africa may also need to be examined more carefully. Some of them may already be importing in all kinds of data bias. Many companies start their AI development by scraping data from social media and so on in the Western World. This web-scraping is not illegal but carries some risks.

---

[47] Journal of Sustainable Development in Africa (Volume 17, No.1, 2015) ISSN: 1520-5509 Clarion University of Pennsylvania, Clarion, Pennsylvania. https://jsd-africa.com/Jsda/V17No1-Spr15A/PDF/African%20Cultural%20Values.Sunday%20Owoniyi.pdf (Retrieved 26/09/2023)
[48] https://www.technologyreview.com/2023/03/28/1070390/what-if-we-could-just-ask-ai-to-be-less-biased/ (Accessed 26/09/2023)

Africa and Africans are some of the least represented people in cyberspace. If you come from Africa, how much of what you see online represents you? So, web-scraping also introduces bias because this scraped data includes various cultural assumptions, including the needs of advertisers on the websites scraped etc.

The world is only just waking up to this problem. Strong Ethics and Governance frameworks can help AI with balancing such biases by asking the right kinds of questions.

Data Rights in Africa

From 20 African countries, the African Digital Rights Network (ADRN) brings together more than fifty activists, journalists, policymakers and researchers, to work on digital rights in Africa[49].

As their names imply, both the African Digital Rights Network is and the African Digital Rights Hub are focused on digital rights in Africa, but from different perspectives.

---

[49] https://www.africandigitalrightsnetwork.org/ (Accessed 26/09/2023)

The ADRN are focussed on digital rights of individuals in a democracy and against illegal surveillance by governments. A collective book by the ADRN group, is titled "Digital Citizenship in Africa: technologies of agency and repression", Edited by Dr. Tony Roberts and Dr. Tanja Bosch with an imprint by Zed Books and published by Bloomsbury.

The Africa Digital Rights Hub (ADRH) is a not-for-profit "think and action tank"[50]. Based in Accra Ghana, the Hub gathers academic researchers, policy makers, regional and international bodies, and other stakeholders, to promote Pan-African research and capacity building on digital rights. In mid-2023, they released a report on the perspectives of young people on digital rights and data privacy in Ghana.

Data Ownership

When we start collecting AI data in Africa, our data safeguarding policies are often still in their infancy. How will the rights of those who participate in AI based projects in Africa be protected? Where does the

---

[50] https://africadigitalrightshub.org/

data collected in Africa go? Who will protect the data rights of ethnic and other minorities within Africa?

Now, who owns that data? Does it belong to those who brought that AI system into the African continent or to the nation and to the individuals where it was deployed? How do we plan for risks of cyberattack? Do we have plans for disaster recovery? All these, and many other questions need to be asked and answered.

Data sovereignty raises further questions. Should African Nations legislate to localise the data, and centre everything about the data in the locality, or do we internationalise it so that it can be used more widely across the continent and internationally? **The answer is that it all depends on how the data is governed.**

Regional bodies like the West Africa Data Protection and Privacy Rights Coalition (WADPPRC) and Network of African Data Protection Authorities (NADPA) are working to harmonize policies across borders. The African Union's (AU) 2022 Data Policy Framework sets out standards for personal data protection and cross-border data flows. While quite promising, these efforts still face limitations in consistency of enforcement and oversight capacity. Rapid technology change is always a challenge and demands that we move faster too.

The Data Protection Act is very fundamental to what we are talking about. Ghana, Kenya, Madagascar, Mauritius, Nigeria, Rwanda, South Africa, Togo, Uganda, and Zimbabwe are all countries that we know have already put in place some of the kinds of legislation and legislation, legislation and regulation that will be needed as they come into the age of AI. There are more below. However, there are many who are yet to successfully begin to legislate data protection.

Data Protection Laws in Northern Africa

The Konrad Adenauer Foundation

In September 2022 Dr. Patricia Boshe and Prof. Dr. Moritz Hennemann, both of the University of Passau, reviewed Data Protection Laws in Northern Africa and Regulatory Approaches, Key Principles, and Selected Documents for the Konrad Adenauer Foundation Think Tank[51].

The report provides an in-depth analysis of the status and trends of data protection laws in 7 North African countries - Algeria, Egypt, Mauritania, Morocco, Libya, Sudan, and Tunisia. It focuses on

---

51

https://www.kas.de/documents/265308/22468903/230406_DataProtection LawsNorthernAfrica_KAS_Web.pdf/ac468c6d-3b82-44d8-bd7e-00ace3906a5b?

regulatory approaches, key principles, and selected legal instruments in these countries.

The analysis is limited to a textual review of data protection statutes and constitutional law ("law in the books"). It looks at the development and status of regional/sub-regional data protection frameworks in Africa and considers political and international influences on the development (or lack of) data protection laws in North Africa.

For the 5 countries with comprehensive laws (Algeria, Egypt, Mauritania, Morocco, Tunisia), it compares alignment with the EU's General Data Protection Regulation (GDPR). The laws in these 5 countries closely resemble the GDPR, but some differences exist like omitting the transparency principle.

Transfer of data rights to heirs is an aspect in some countries' laws not explicitly in EU framework. Despite similarities, none of the 5 countries have received an EU adequacy decision recognizing their law as equivalent. Enforcement challenges exist across the region. Private enforcement also seems limited currently.

The report provides a useful legal analysis of the status of data protection in North Africa, with insights into trends, influences, principles, and recommendations. The report recommends promoting cooperation between data protection authorities in the region, and with legislators, to work towards harmonized frameworks.

The African Union

The African Union (AU) can play a bigger role in the development of AI. Clearer pan-African guidelines and norms are needed around data labelling, consent, localization and sharing between public-private entities.

The AU could expedite wider adoption of its AI governance frameworks[52]. This can help align standards for transparency, accountability, and fair use of AI. Member states should collaborate to institute regional regulatory bodies that can oversee and approve the African AI systems.

To incorporate African values and aspirations into a truly African AI, the AU can help encourage the involvement of non-technical voices. Building an African AI needs the voices of the community, or wider society as much as it needs the voices of industry. It needs all stakeholders, including faith leaders, their congregations and more.

Private sector firms can volunteer self-regulation and codes of conduct as part of Environmental, Social and Governance (ESG) mandates. Open data access while still preserving privacy will help AI innovation.

---

[52] https://au.int/en/treaties/african-union-convention-cyber-security-and-personal-data-protection

More training programs for policymakers and technologists on African values and ownership are needed. Urgent action is required.

Data Gathering in Africa
Large sets of data are the secret to AI development. Africa by population deserves to become another AI powerhouse. Africa also has the youngest population globally with a median age under 20 years old. Africa is larger than the USA, Europe, and Russia combined. That means that Africa is ripe for a huge technology leap forward.

Africa's population is roughly equal to India or China (1.4 bn). Africa's population is projected to reach 2.5 billion by 2050, which will mean that 1 in 4 people on Earth will be African.

But a lack of standard data methods makes trans-continental AI and machine learning difficult. The African Union (AU) Data Policy Framework is designed to lead to the creation of the African Digital Single Market (DSM). However, the Framework does not define data classification. It only encourages categorizing personal and non-personal data. It advocates giving individuals and organizations control over their data.

It cautions against broad data localization requirements. But there are needs to quickly resolve tensions between data localisation and data sovereignty as against the internationalisation of data for continental AI development.

The African Union will need to encourage agreement on a Master Data Management System, urgently, to prevent multiple schemes of data categorisation across the continent. Robust data protection and privacy mechanisms aligned across Africa are to be encouraged, as well as open data access and cross-border data flows.

AU states should collaborate through regional institutions on governance and standards to exert leadership globally. A 5-phase implementation plan engages stakeholders. Overall, the Framework seeks to harmonize data policies across Africa and position the continent as a standard-maker in global data regulation conversations.

Africa, the Young Continent

CHAPTER 5- DEVELOPING AI TALENT AND SKILLS

The Core Idea in Chapter 5:
More strategic investment must be made by African into developing AI skills and talent in Africa.

Harnessing the full potential of AI to drive economic and social progress will require major investments in developing homegrown talent. Currently, Africa faces a significant shortage of technical capabilities for cutting-edge research and development in AI. A 2019 study found only 22% of African universities offered dedicated courses in AI or machine learning.

African and International Skills Gaps

In "Foundational skills for a more inclusive Fourth Industrial Revolution (4IR) in Africa" by Louise Fox and Landry Signé, the Brookings Institute noted that while the number of young African professionals in the IT sector is growing rapidly, the sector still only employs less than 1% of the African labour force.

The global expansion of AI technology has revealed the importance of skills development for secondary and tertiary graduates, another area where African education systems need substantial improvements[53].

In 2020 in the European Union, ICT specialists made up 3.9% of the workforce[54]. In 2021 (post-Brexit) the UK, had 1.5 million people in professional IT and telecoms roles, making them 4.3% of the workforce[55]. In the same year (2021), in the USA, computing occupations accounted for 3.3% of the workforce (around 5 million jobs)[56].

50 million people in India are estimated to have jobs in the IT/ITeS sector but the estimates differ between 3% to 10% depending on who you read[57][58]. But the IT employment in the UK, Europe, USA and India is significantly higher than 1% in Africa.

---

[53] https://www.brookings.edu/articles/foundational-skills-for-a-more-inclusive-fourth-industrial-revolution-4ir-in-africa/ (Accessed 24/09/2023)
[54] https://ec.europa.eu/eurostat/statistics-explained/index.php?title=ICT_specialists_in_employment (Accessed 24/09/2023)
[55] https://www.ons.gov.uk/employmentandlabourmarket/peopleinwork/employmentandemployeetypes/bulletins/employmentintheuk/august2021 (Accessed 24/09/2023)
[56] https://www.bls.gov/emp/tables/emp-by-detailed-occupation.htm (Accessed 24/09/2023)
[57] https://www.ibef.org/industry/information-technology-india.aspx (Accessed 24/09/2023)
[58] https://nasscom.in/knowledge-center/publications/future-jobs-2022 (Accessed 24/09/2023)

ICT specialists in the EU grew by 57.8 %, from 2012 to 2022, almost 7 times as much as the overall expansion of employment (8.8 %). Some 81.1 % of ICT specialists in the EU in 2022 were men, and only 18.9 % of women. In that year, 65.4 % of ICT specialists in the EU had finished a tertiary level course of education or training[59] [60].

African AI Skills Growth

However, Africa has great potential to grow its digital workforce and IT sector as technology adoption expands. Much depends on what Africa does with this opportunity. Several government initiatives are underway to build capabilities:

As we have seen, South Africa's new AI institute and university research centres target high-level R&D skills. Recognising that AI is going to create new jobs that require new technical skills, these and other projects across South Africa have also focussed a lot of resources on skills for the youth. This is also happening in Ghana and other places in Africa. While upskilling the youth in this AI technology, we should also produce those who will provide jobs in the future. Leadership and management skills are needed in these fields as much as in any other.

African Based Degrees in AI

Here are some universities in Africa that offer degree programs related to Artificial Intelligence:

- University of Cape Town, South Africa, MSc in Artificial Intelligence, 2 years

---

[59] https://ec.europa.eu/eurostat/statistics-explained (Accessed 24/09/2023)
[60] https://www.statista.com/statistics/284968/it-software-and-computer-services-economy-employment-in-the-united-kingdom-uk/#:~:text=As%20of%20September%202022%2C%20over,thousand%20people%20were%20self%2Demployed.

- University of the Witwatersrand, South Africa, BSc in Data Science and Computer Science, 3 years
- University of Pretoria, South Africa, BEng in Computer Engineering with focus on AI, 4 years
- Stellenbosch University, South Africa, MSc in Artificial Intelligence, 1 year
- University of Ghana, MSc in Artificial Intelligence, 2 years
- Ashesi University, Ghana, BSc in Computer Science with Machine Learning courses, 4 years
- University of Lagos, Nigeria, MSc in Artificial Intelligence and Data Science, 2 years

Some other notable universities offering AI programs in Africa include Makerere University (Uganda), University of Nairobi (Kenya), University of Mauritius, and University of Rwanda. The programs listed cover a range of undergraduate and postgraduate degrees focused on various aspects of AI like machine learning, data science, deep learning, robotics, etc.

- Makerere University, Uganda, BSc in Computer Science with AI electives, BSc in Software Engineering with AI electives, 4 years.
- University of Nairobi, Kenya, MSc in Computer Science with specialization in AI, 2 years.
- University of Mauritius, BEng in Artificial Intelligence and Robotics, 4 years, Bachelor of Engineering.
- University of Rwanda, MSc in Computer Science with specialization in AI, 2 years.
- Addis Ababa University, Ethiopia, MSc in Computer Science with AI specialization, 2 years.

- Ahmadu Bello University, Nigeria, BSc in Computer Science with AI electives, 4 years.
- Université Félix Houphouët-Boigny, Ivory Coast, Master in Artificial Intelligence and Business Intelligence, 2 years.

Universities have partnered with IBM and Huawei on curriculum design. But more strategic alignment between education programs and labour market needs is required to develop pathways from school to employment.

Nigeria's Youth for Technology Foundation collaborates with universities on AI skills training for youth. Rwanda has introduced coding and robotics clubs in thousands of schools to build foundational skills. There is also the AI for Development (AI4D) program in Egypt

Here are more details on the AI/Data Science Master's Programs at three African universities:

MSc in Data Science and AI at the University of Ghana

- Two-year graduate program launched in 2020 at the university's Institute of Applied Science and Technology.
- Focuses on areas like machine learning, neural networks, computer vision, natural language processing.
- Emphasizes hands-on experience through lab sessions, industry projects, internships.
- Aims to develop students' skills to apply AI/data science to sectors like healthcare, agriculture, finance.
- Supported by organizations like Mastercard Foundation, IBM, and Vodafone Ghana

AI for Development (AI4D) program in Egypt

- Joint program between Egypt's Information Technology Institute and IBM launched in 2019.
- Provides graduate students with access to resources at IBM research labs.
- Curriculum covers areas like machine learning, visual recognition, neural networks, and ethics.
- Students work on projects tackling challenges in healthcare, transportation, energy, and agriculture.
- Goal is to build local capacity in AI to drive economic growth and social progress.

Artificial Intelligence and Data Science (AID) program at the University of Lagos in Nigeria

- Two-year master's program hosted by the university's School of Computer Science.

- Blends computer science, statistics, applied mathematics, and domain knowledge.
- Capstone projects and internships provide practical experience for students.
- Supported by partnerships with tech companies like Oracle, IBM, and Microsoft.
- Aims to develop Nigeria's capabilities in AI/data science across different industries.

Across Africa and Beyond

Some of these programmes are also continental and global. That is, these universities have links in other countries across Africa and around the world. Universities are increasingly part of larger partnerships. Industries are also increasing partnering with education to build a workforce that is fit for purpose.

There are scholarship programmes, internships, and many other incentives. So if you want to be part of AI, a lot of doors have been opened for you. There are tax rebates for companies, vocational training schools also being incentivized to add AI classes to their curriculum. All this is creating a critical mass within the population. The skills, the manpower, and the level of strategic thinking, in AI across Africa will change because this training is increasingly available from the grass roots up to postgraduate levels in the universities.

The private sector and other agencies also play a key role in raising the skills for AI across Africa. All of this needs to grow faster.

Egyptian Startups Accelerator Programme
Graduates of 23 Egyptian startups were celebrated at the third group of the 3-month long Egyptian startups Accelerator Programme [61]. They were supported by the United States Agency for International Development (USAID) in partnership with the Information Technology Industry Development Agency (ITIDA), the National Bank of Egypt (NBE) and Egypt Post.

During their training, the Egyptian entrepreneurs were given networking opportunities, mentorship, and other resources. Topics tackled included AgriTech, the circular economy in Egypt, the Egyptian venture capital space, and the impact of AI on revolutionizing businesses, lending in the B2B space, and market trends.

Google in Ghana Goes Continental
Train 100,000 AI experts across Africa and build startups is another aim of Google's AI centre in Ghana[62]. Research teams from around Arica

---

[61] https://www.zawya.com/en/press-release/companies-news/usaid-fuels-entrepreneurial-growth-in-egypt-la3rp7bv (Accessed 27/09/2023)
[62] https://research.google/locations/accra/ (accessed 230923)

and across the world partner with the Google Research teams in Accra to meet these goals and tackle global challenges. Their focus on theoretical and applied artificial intelligence allows partnerships that pilot many initiatives of particular interest to Africa in the area of sustainability.

African Robotics Network (AFRON)
Shared interest in robotics in Africa brings individuals, institutions, and organizations together in the African Robotics Network (AFRON). The network seeks to improve robotics-related partnerships through better communication. Their work in robotics includes automation, computer vision, machine learning, and signal processing,

AFRON provides educational, research, and industry-oriented projects, events and other meetings across the continent. They are also present at international events.

Pan-African Robotics Competition
From across Africa and its diaspora PARC brings together robotics teams from middle and high schools for a youth robotics competition every year[63]. Challenges for competing teams are in three categories, based on real-world topics related to the sustainable development of Africa and relevant to science, engineering, design and performance. Prizes include scholarships.

In 2023, PARC will host the Pan-African Science and Engineering (PACE) for the first time. This event showcases research, projects, and inventions of the young participants in science, technology, engineering, and mathematics (STEM) from all over Africa. PACE is sponsored by Google.

---

[63] https://parcrobotics.org/ (accessed 230923)

Youth vocational training is equally important to build operational capabilities in AI model development, data analysis and machine learning engineering. Curriculum reform should introduce AI modules across all fields from agriculture and healthcare to the arts given its crosscutting impacts.

International partnerships can provide valuable knowledge exchange opportunities. Ultimately a culture of lifelong learning must take root to continually upgrade skills.

With foresight and investment, Africa's youthful demographics can become a competitive edge. Developing high-caliber AI talent will drive innovation in local contexts while helping address ethical application and inclusion gaps. Africa needs to believe in its youth and equip them technologically for the future.

CHAPTER 6- WHAT NEXT?

The Core Ideas in Chapter 6:
African AI progress in the development of the micro-entrepreneurial culture. Advice on prospective entrepreneurial opportunities, skills needed, and resources available to succeed in African AI startups.

The mobile data revolution across Africa is perhaps the most important part of the growing AI revolution, enabling digital access for entire populations. In turn, AI looks set to have an even greater impact on Africa than the arrival of mobile telephony.

In her 2018 Issue Brief, for The Atlantic Centre Africa Council titled Coming to Life - Artificial Intelligence in Africa, Aleksandra Gadzala reviewed the state of artificial intelligence (AI) adoption at that time[64]. She noted then, that while excitement existed about AI's potential, progress had been limited except in a few countries like Kenya, South Africa, Nigeria, Ghana, and Ethiopia. That has since improved as this my book notes in several places.

Importantly, back in 2018, Gadzala provides examples of early AI applications in agriculture, healthcare, financial services, education, and other sectors and highlighted several factors enabling AI success stories in the leading African nations, including:

- Existing digital foundations via widespread mobile phone usage and existing mobile tech solutions that AI can build on.

---

[64] https://www.atlanticcouncil.org/wp-content/uploads/2019/09/Coming-to-Life-Artificial-Intelligence-in-Africa.pdf

- Supportive government policies and national prioritization of technology innovation.
- Emerging cultures of tech research and innovation cantered around universities and start-up hubs.
- Global partnerships providing financing, expertise, and technological capabilities.

Today (2023) availability of mobile phone coverage, and knowledge of how to use available online platforms is creating a new generation of micro-entrepreneurs across many parts of Africa.

The Micro Entrepreneurs

Strong adoption of mobile services is also driving the expanding digital economy in Economic Community of Central African States (ECCAS). This has broken new grounds in job creation and access to services for women and young people.

E-commerce platforms are becoming popular with a new generation of entrepreneurs. According to the United Nations Conference on Trade and Development (UNCTAD) the informal channels that carried 90% of transactions across Sub-Saharan Africa are being challenged and changed by the growing access to ecommerce, mainly through mobile phones[65].

Political and business leaders and other policy makers should help ensure local communities have affordable power supplies for recharging batteries, realistic network tariffs and also remove other barriers that can create a digital usage gap.

Writing for the Carnegie Endowment, in an article titled "To Close Africa's Digital Divide, Policy Must Address the Usage Gap", Jane Munga points out that there can be adequate coverage in an area, but

---

[65] https://data.gsmaintelligence.com/research/research/research-2021/enabling-e-commerce-in-central-africa

if people there are unable to utilise that coverage, there is what can be called a "Usage Gap"[66].

Mastercard Foundation report on Micro-entrepreneurs in a platform researched the e-platform practices of 27 micro-entrepreneurs in Kenya[67]. The research focussed on their use of social media, e-commerce sites, and online freelancing platforms. The report finds that the micro-entrepreneurs use a mix of platforms like Facebook, WhatsApp, and Instagram to connect with customers. many transactions still happen offline. There is frequent interplay between online ("tech") and offline ("touch") interactions in business processes.

---

[66] https://carnegieendowment.org/2022/04/26/to-close-africa-s-digital-divide-policy-must-address-usage-gap-pub-86959 (Accessed 23/09/2023)
[67] https://mse.financedigitalafrica.org (Accessed 23/09/2023

The more digitally savvy micro-entrepreneurs use e-commerce platforms like Jumia and OLX to reach more customers. WhatsApp groups are emerging as "digital trade unions" where micro-entrepreneurs in the same sector share information. YouTube, Facebook groups, and offline connections are used for skills training and advice.

Some micro-entrepreneurs do online freelancing (e.g., writing) in the evenings as a "side hustle." They adapt platforms in creative ways to access emerging opportunities. Ratings and reviews remain very important for credibility on e-commerce and freelancing platforms.

The Mastercard report does highlight possible investment opportunities in areas like credibility systems, transformational training, and integrating payments into messaging apps. There are also possible areas for future research on topics like blending tech/touch, the processes of transforming side hustles into mainstream livelihoods, and protecting vulnerable online workers.

M-Kopa in Kenya

The National UAE based news website reported in 2018 that the Kenyan Silicon Savannah flagship technology ecosystem, including the planned Konza Technology City had become stranded. But the social enterprise M-kopa which was born out of the original Konza Technology City, is thriving, and provides electricity for 300,000 customers or 5% of Kenyan homes[68].

M-Kopa keeps mobile phones charged, giving people access to e-commerce and the Internet, with huge knock-on effects on livelihoods and lifestyles, enabling the start of the AI revolution at the grass roots.

---

[68] https://www.triplepundit.com/story/2016/how-solar-startup-became-multinational-social-enterprise/29001 (Accessed 25/09/2023)

THINK BIG!

In the next few pages, we will be looking at building your own AI start up. Please just remember that because you may start small does not mean that you have to always "think small".

Wherever you start from, THINK BIG. Always think bigger than where you are. Always dream, think and plan for the next step, and always try to make your next step your best step.

Here are some examples of successful African AI startups, their offerings, market size, and funding attracted.

- **Aerobotics (South Africa)** - Uses AI and drone imagery to provide analytics for farmers. Serves over 300,000 hectares globally. Raised $17 million in funding.
- **Voyc.ai (South Africa)** - Offers an AI-powered customer engagement platform. Serving large telcos and banks in South Africa. Raised $5.5 million in funding.
- **MDaaS Global (Nigeria)** - Uses AI for diagnostic imaging and analysis. Active in 9 African countries. Raised $2.3 million in funding.
- **Swvl (Egypt)** - An AI-powered transportation startup providing ride-hailing and shuttle services. Operating in 7 countries globally. Raised over $100 million in funding.

African AI startups are attracting investment and gaining market share in sectors like agriculture, healthcare, education, transportation, and more.

Starting Your Own AI Business

If you want to start a line of business, gain an education, determine policy in Artificial Intelligence, or if you are just curious, this book is for you. If you are connected with or want to learn more about Artificial Intelligence in Africa, then this is a book you cannot afford to miss.

upwork

The Team at Upwork "the world's work marketplace" offers you some advice on starting your own AI business. Here are their top tips. For the full details of the article, please follow the link in the footnotes below.[69]

1. Understand artificial intelligence and its applications

2. Define your business model

3. Build a strong team

4. Leverage AI technology and infrastructure

5. Develop AI products and solutions

6. Craft marketing and business development strategies

7. Scale and grow

Machine Learning Project Readiness Tool

Amelia Taylor (PhD) is a lecturer in Artificial Intelligence, Computational Intelligence and programming modules at the University of Business and Applied Sciences (MUBAS), Malawi, and is

---

[69] https://www.upwork.com/en-gb/resources/ai-startup (Accessed 27/09/2023)

also the main organiser of machine learning conference in Malawi IndabaX Malawi[70].

In her blog for the Deep learning Indaba, on Tue 11. Apr 2023 titled "ML-READINESS", Dr Taylor has proposed developing a ML-readiness tool along the lines of the Technology Readiness Levels (TRLs). These are used for consistency in considering and assessing the collective maturity of varieties of technology during corporate mergers and acquisitions[71].

TRLs are usually part of a technology readiness assessment (TRA) of the technology requirements, the strategic plans and operational business models, as well as the previous track record of the existing and proposed technology[72].

In her blog post, she emphasises the need for you to test business model readiness when venturing into a ML startup and using a tool to measure your ML-readiness level in the following areas.

General components of ML readiness:

- Reliable data storage
- Processing power
- Data integration capabilities
- Access to skills and technology and availability of training)
- Datasets and ML Libraries for Design
- Costs and Funding
- Ability to develop scalable/marketable prototypes

---

[70] https://mwdata.science/speakers/amelia/ (Accessed 24/09/2023)
[71] https://deeplearningindaba.com/blog/2023/04/ml-readiness-level/ (Accessed 27/09/2023)
[72] Mihaly, Heder (September 2017). "From NASA to EU: the evolution of the TRL scale in Public Sector Innovation" (PDF). The Innovation Journal. 22: 1–23

- Tools for quality/risk management
- Policy environment
- Collaboration opportunities

To succeed, every business venture must provide sufficient value to its clients at reasonable costs, quality and profit. As many tech startups have discovered enthusiasm and even technical expertise are often not enough to ensure your success.

Using simple tools like the ones provided in this chapter of the book will help reduce the uncertainties in the early phases and during times of change in your business. They will help to give you a sober practical assessment of the technology, people and processes and products that you are assembling for your project.

These kinds of assessments (and there are many more out there) will help you sharpen your strategies and clarify your roadmap.

Emerging AI Apps

Here are some of the emerging uses Ai Apps that might help to inspire you. Avijeet Biswal, a Business Analyst wrote an online article for digital skills training SimplyLearn titled "AI Applications: Top 18 Artificial Intelligence Applications in 2023".

In the Article, Avijeet lists 18 sectors of growing influence for Artificial Intelligence Applications [73].

The AI applications are working across:

- E-commerce
- Education
- Lifestyle,

---

[73] https://www.simplilearn.com/tutorials/artificial-intelligence-tutorial/artificial-intelligence-applications (Accessed 25/09/2023)

- Navigation
- Robotics
- HR
- Healthcare
- Agriculture
- Gaming
- Automobiles
- Social media
- Marketing
- Chatbots
- Finance
- Astronomy
- Data security
- Travel/transport
- Automotive industry.

In e-commerce, AI enables personalized recommendations, virtual shopping assistants, and fraud prevention. In education, AI assists with administrative tasks, creates smart content, powers voice assistants, and enables personalized learning.

For lifestyle, AI applications include autonomous vehicles, spam filters, facial recognition, and recommendation systems. AI also improves navigation through lane detection, traffic analysis, and route optimization by companies like Uber.

Robotics leverages AI for sensing, planning, inventory management, and human-like interactions. HR uses AI for blind hiring, candidate screening, and modelling the talent pool.

Healthcare applies AI for disease detection, diagnosis, drug discovery, and analyzing medical data. AI in agriculture helps identify soil defects, weeds, and assists in harvesting. It creates human-like NPCs in gaming.

Social media sites use AI for personalized feeds, content moderation, fraud detection, and recommendations. Marketing employs AI for targeted ads, content creation, campaign analysis, chatbots, and personalization.

Finance leverages AI for wealth management chatbots, fraud detection, risk assessment, and task automation. Astronomy uses AI for galaxy analysis, signals detection, exoplanet identification, and handling large datasets.

AI strengthens data security through threat detection, vulnerability identification, prevention, and quick response. It optimizes travel and transport via route planning, traffic management, ride-sharing coordination, and logistics.

Avijeet's article provides a broad overview of the diverse applications of AI across many industries and sectors. Where will your Ai journey take you?

Key skills for AI/Machine Learning Entrepreneurs:

Technical Skills:
- Proficiency in programming languages like Python, R, Java, C/C++ etc.
- Knowledge of ML frameworks like TensorFlow, PyTorch, Keras etc.

- Understanding of ML algorithms like regression, classification, clustering etc.
- Ability to build and optimize ML models
- Knowledge of math - linear algebra, calculus, statistics
- Data analysis and visualization skills
- Cloud computing skills to scale models

Business Skills:
- Strong problem solving and analytical abilities
- Ability to identify real-world problems that ML can solve
- Understanding of target industry and market
- Product design and management skills
- Marketing and business development skills
- Fundraising and investment pitch abilities
- Project management and product development process skills

Soft Skills:
- Creativity and innovation
- Communication and storytelling abilities
- Collaboration skills to work across teams
- Leadership and team building qualities
- Persistence and learning mindset

A mix of technical prowess in ML and coding, business acumen, and soft skills like creativity and communication is important for being successful with an AI/ML startup. The blend of skills depends on the specific roles and needs.

Funding Application Criteria

We have listed many sources of funding already in this book. Each will have their own criteria and you will need to look carefully at what is required. However just to give you an example, if you want support from the African Development Bank, here is a summary of the main points for the eligibility criteria for AfDB private sector funding[74]:

- Project must be in an African Development Bank member country.
- Enterprise must be majority privately owned or publicly owned with financial autonomy.
- Funding is for establishing, expanding, diversifying, or modernizing productive enterprises (capital expenditures).
- African Development Bank participation cannot exceed 33% of total project cost, usually a minimum of $3 million.
- Evidence required of strong integrity, reputation, and financial standing of enterprise.
- Proposals screened for eligibility first then sent to operational teams for review.
- Responses provided within 15 days of submitting proposal.

---

[74] https://www.afdb.org/en/sectors/private-sector/how-work-us/funding-request#:~:text=The%20enterprise%2Fproject%20must%20be,productive%20enterprises%20(i.e.%20CAPEX)%3B (Accessed 26/09/2023)

The key criteria are that the project must be in an AfDB member country, have significant private sector involvement, be for capital expenditures to develop enterprise capacity, limit AfDB funding to a minority share, and demonstrate strong financial and ethical standing. The process involves initial eligibility screening then more detailed review by operations teams. The AfDB application forms are at this link[75].

Email your completed forms to: PrivateSectorHelpDesk@afdb.org

---

[75] https://www.afdb.org/sites/default/files/2020/05/04/nso_funding_request_contact_form_en.pdf

CHAPTER 7- CONCLUSIONS

This short book introduces AI and its potential benefits for Africa's development as well as some potential pitfalls. It discusses current AI investments and projects in Africa through case studies of AI applications for social good, particularly in healthcare.

The book emphasizes the urgent need for strategic investment in developing homegrown AI skills and talent in Africa to create inclusive, responsible AI ecosystems.

Building AI skills can create new technical jobs and lead to socio-economic development via African AI startups that tackle local challenges. However, the book lacks diverse African perspectives. It could be enriched by more insights from African AI experts, entrepreneurs, and youth. More of this needs to become available.

The book aims to motivate and guide different audiences on AI's opportunities. For aspiring tech entrepreneurs and students, it explains the AI landscape and how to start impactful businesses. For policymakers, it advocates investing more in skills development. For business leaders, it encourages research and startups after demonstrating successful use cases.

The book covers relevant topics like promoting responsible AI governance and ethics in Africa alongside skills development. While remarkable progress has occurred via African micro-entrepreneurship, prospective entrepreneurs need to recognize opportunities, develop requisite skills, and leverage available resources to succeed. Overall, strategic skills investment and inclusive, responsible AI ecosystems are critical for Africa to realize AI's long-term benefits.

Across Africa, AI is already catalysing transformation in fields as diverse as healthcare, agriculture, education and finance. This book has analysed applications, policies, and partnerships needed to maximize these promising beginnings for economic and social development.

So where are we? Do we use AI or do we lose out? I think the consensus must be that we need to use AI. Africa needs to jump in with both feet to grab AI with both hands and use it. We need to use AI to transform our old industries, while creating new ones.

We need to retrain our existing workforce for the emerging new world of AI. We need to really engage with the young people to see how AI can help create jobs for them. AI is presenting us with a huge opportunity, greater than the arrival of mobile telephony. We must use it well.

Karen Hao of the MIT Technology Review believes that the future of AI research is in Africa[76]. The Economist also says that Africa is the only region in the world where venture capital is not slowing down[77]. But Annie Njanja, and Tage Kene-Okafor of TechCrunch Market Analysis predict a slowdown in the flow of venture funding to Africa in 2023[78].

The signs for now are that AI in Africa is still growing.

- Increased investment - AI startup funding in Africa grew from $93 million in 2020 to $134 million in 2021, with high profile investors like Google, IBM, and Microsoft involved.
- Government initiatives - Countries like Kenya, South Africa, and Egypt have launched national AI strategies and research institutes to promote AI development.

---

[76] https://www.technologyreview.com/2019/06/21/134820/ai-africa-machine-learning-ibm-google/ (Accessed 27/09/2023)
[77] https://www.economist.com/middle-east-and-africa/2022/07/21/african-startups-are-raising-unprecedented-amounts-what-next (Accessed 27/09/2023)
[78] https://techcrunch.com/2023/01/13/africa-predicted-to-experience-sustained-funding-slowdown-in-2023/ (Accessed 27/09/2023)

- Talent development - Universities across Africa are creating undergraduate and graduate programs in AI and machine learning to build a skilled workforce.
- Entrepreneurship - There has been a surge of AI startups applying AI across sectors like healthcare, agriculture, transportation, finance, and education.
- Adoption in key industries - Major telecom, banking, and natural resource companies in Africa are adopting AI solutions for things like customer service, fraud prevention, and operational efficiency.
- International partnerships - Multinational collaborations like the IBM Research Lab in Kenya and Google AI center in Ghana connect African talent to global AI networks.
- Improving infrastructure - Government and private sector investments to expand internet access, data centers, cloud computing and other AI-enabling infrastructure.
- Applications for development - AI is being applied to support development in areas like disease diagnosis, water management, gender inclusion, wildlife conservation and more.
- Emerging regional frameworks - The African Union and regional bodies are working on AI strategies, data protection laws, and other frameworks to enable responsible AI growth.

Karen in Egypt is an example of robotics, producing industrial robots. In Ghana we have seen mPharma and there is also Zipline there, which is also in Rwanda.

We saw a lot of developments in South Africa. In Nigeria, Hello Tractor is using AI in scheduling of the equipment. In the coming years, Africa will become more of an exporter of AI.

AI can help leapfrog development challenges from literacy to inclusive credit when grounded in Africa's realities. Public and private collaboration is crucial. More investment in AI is needed.

One of the major problems is that governments all over the world are struggling to keep pace with technology. Thoughtful governance and policies are urgently required to address risks around data bias, job losses, privacy violations before AI systems become entrenched.

- Developing competitive homegrown talent through revamped education and training will drive innovation and opportunities.
- Balanced partnerships between government stewardship and private sector dynamism to build equitable, ethical AI ecosystems.

Ultimately AI should reflect African values of community, transparency, and uplifting human dignity rather than concentrating power.

There is growth in investment, talent, startups, infrastructure, and technology and policy adoption. All these are pointing to Africa's momentum in developing and harnessing AI to address local needs and opportunities.

Looking ahead, the coming decade will be decisive for Africa to lead in shaping AI as a force for empowerment and shared prosperity rather than division. If stakeholders act swiftly and jointly guided by foresight, ethics and determination to forge an African path, the "AI Age" can usher in historic social and economic flourishing.

Local innovation in African AI solutions can build new export industries around technology services. Data analytics and AI modelling can support better policy, regulation, and governance to catalyse economic growth. New AI startups or partnerships are springing up every year. African Governments must move faster to keep pace.

For all the positive potential, AI comes laden with significant risks that require mitigation. If transparency and fairness are not built into algorithms, machine learning models can perpetuate and exacerbate societal biases and discrimination. Imported systems may embed cultural viewpoints and values misaligned with Africa. The collection of data to train AI models also raises privacy concerns and cybersecurity vulnerabilities without proper governance.

While AI will generate new high-skilled jobs, lower-skilled repetitive work will face displacement through automation. A UK Royal Society report estimates that in Ghana alone over 11% of current jobs are at risk from robotics and intelligent systems. Without retraining programs, unemployment and inequality could surge.

Thus, responsible, and ethical AI development remains critical. Principles like accountability, transparency (explainability) and inclusiveness must be core design values, not afterthoughts. Policymakers have a key role through governance frameworks that incentivize innovation while managing risks proactively. Strong partnerships between government, private sector and civil society will be vital.

Regional Cooperation
Realizing the full benefits of AI for economic and social development in Africa will require active leadership and collaboration between government and private enterprise. On the public sector side, initiatives like Ghana's ICT-enabled Accelerated Development Strategy demonstrate high-level prioritization. Supra-national bodies like the African Union and AfDB have developed policy frameworks to coordinate AI growth.

Governments play a vital role through creating a policy environment for infrastructure. They often also provide research funding, incentives for technology firms to invest, and education policies to develop digitally skilled workforces.

As we have seen, Governments also shape legal and ethical guardrails for AI through data protection laws and governance institutions that uphold transparency and accountability. However, many nations still lack comprehensive strategies and institutions to support AI development.

Private companies like Microsoft, Huawei, Google and IBM are providing critical infrastructure, R&D and skills training across the continent. But concerns persist around misalignment with local needs, concentrating power and profits, and importing negative biases. There are also concerns around the export of data from AI and other digital systems operating in Africa, including the alleged Chinese data for technology swaps.

Greater incentives and contractual requirements for technology transfer and local partnership are needed to build equitable innovation ecosystems.

A balanced approach should leverage the dynamism and expertise of the private sector while ensuring government stewardship and inclusive development. Strategic policies and joint forums can align priorities and incentives. Ultimately, a networked system with nodes across public agencies, private firms, academia and civil society is needed for participative decision-making on Africa's AI future.

As connectivity plays a central role in enabling e-commerce, AI and building communities, the telecommunications network must also be priority.

Artificial Intelligence in Africa

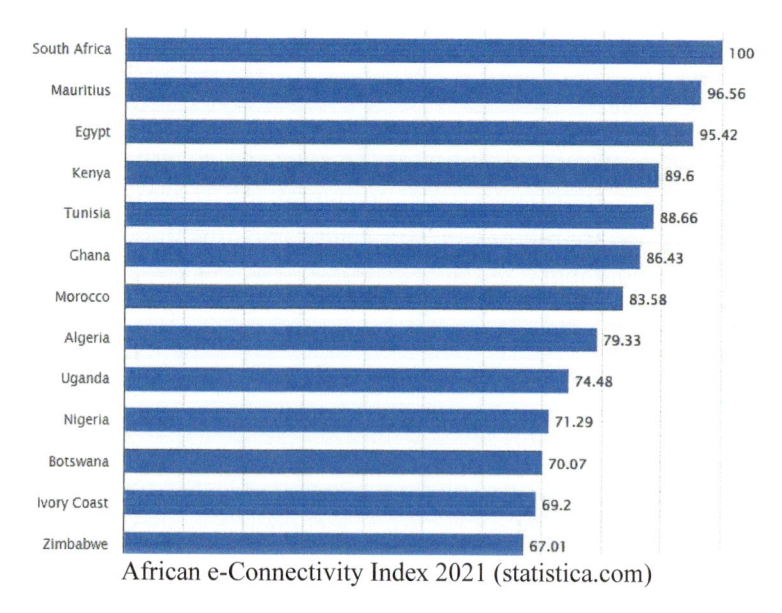

Country	Index
South Africa	100
Mauritius	96.56
Egypt	95.42
Kenya	89.6
Tunisia	88.66
Ghana	86.43
Morocco	83.58
Algeria	79.33
Uganda	74.48
Nigeria	71.29
Botswana	70.07
Ivory Coast	69.2
Zimbabwe	67.01

African e-Connectivity Index 2021 (statistica.com)

So there'll be a need for much more regional cooperation, as we have noted earlier, Pan-African and regional bodies have a lot of work to do. We should not be left behind when the world moves on with AI.

Africa today (2023) is larger than the U S A Europe and Russia combined in population and one in four people will soon be on earth, will soon be Africans. But how do we utilise this? Let's remind ourselves. A big population is a source of Big Data, which is also the secret behind AI. Big data is not everything. How we handle the data is important, but developing Ai and ML without Big Data is difficult.

If you look at the giants of AI around the world, they are people or they are peoples or countries or regions that have huge populations. You think of China, you think of the United States to some extent the European Union. Why is this? Because if you have the huge populations, you can gather huge amounts of data.

That means that you can quickly train AI systems to deal with the issues of that population and you can even use some of the output of that to export your AI product to other parts of the world.

To do that, you have to, to standardise or harmonize your data collection. The African Union Data Policy framework, which is meant to lead to the creation of the African digital single market DSM did not even define data classification. The AU have encouraged categorising personal and non-personal data, but they have not helped by giving a description or a framework to help nations to harmonise their data classification. That really needs to be corrected if it has not been already.

Regional cooperation in AI projects across Africa will be easier when data gathering systems and protocols are harmonized. A model that is like a master data management system needs to be developed for the continent and be flexible enough that nations can adapt it. Should African states model data laws along European of US laws or develop their own models that can fit with both? Africa needs to think through.

There's a window of opportunity for us to act quickly and we need to support the development of AI products trained on African data designed ethically with Africans in control. We need the whole society to be involved as stakeholders and in Africa, the land of faith, that must also include the faith-based institutions.

End Note

Africa is a big place. We could not cover everything and everyone in AI on that vast continent. Apologies if we have not mentioned your favourite AI project. But I hope you have learned, from the start up, or the mature AI project or the policies of government.

What does the AI journey of those in this book tell you about your own roadmap? Who have they partnered with? Who has funded them?

Each fact, figure and story is here to help you and me think about our own possibilities.

This book has introduced AI and its potential benefits for Africa's development. We have discussed current AI investment and projects emerging in Africa, including case studies showing AI applications for social good.

We have covered the need for responsible AI governance and ethics and also emphasized building AI skills and talent in Africa. There has been advice for prospective African AI entrepreneurs.

Perhaps the most important point, is the need to strategically invest in developing homegrown AI skills and talent in Africa.

The book has been mainly aimed at aspiring African technology entrepreneurs, students and others interested in AI, policymakers, and corporations exploring AI opportunities in Africa.

We hope it has inspired you pursue education and careers in AI. If you are a policy maker to invest more in AI skills development. For leaders in business, to increase AI research and startups in Africa, after reading our examples and case studies.

AI entrepreneurs may now better understand the Ai landscape and how to start AI businesses that will tackle local challenges and make a profound impact. This can also lead to socio-economic development.

I have tried to keep the writing clear and accessible to a broad audience, using case studies and examples to help make concepts real. I do wish that I had more perspectives from African AI experts, entrepreneurs, and youth to help make it more insightful.

The book has covered very relevant topics and is designed to motivate strategic AI skills development in Africa and promote responsible governance - which could have significant long-term benefits if achieved. More diverse African voices and balanced perspectives could enrich the analysis.

AI offers major opportunities for Africa's development but more strategic investment in skills, data, and responsible governance is urgently required. AI investment is surging in Africa from global tech companies and multinational agencies, but needs to increase, and responsible governance frameworks are needed to create truly inclusive AI ecosystems.

Case studies prove the success of AI applied for social good in African healthcare. Investment in AI skills can help create new technical jobs across the workforce. Responsible and contextualised AI governance must be built into the future of AI in Africa

Finally, we must repeat that more strategic investment must be made by Africans into developing AI skills and talent in Africa.

AI's progress in Africa has been remarkable, especially in the development of the micro-entrepreneurial culture. But prospective entrepreneurs need to recognise opportunities, develop skills needed, and harness resources available to succeed with African AI startups.

Suggested Further Reading:

Damian Okaibedi Eke, (Ed.), Kutoma Wakunuma, (Ed.), Simisola Akintoye(Ed.), Responsible AI in Africa Challenges and Opportunities Pub:Palmgrave Macmillan, 2023. https://link.springer.com/book/10.1007/978-3-031-08215-3

Brokensha S., (Auth), Kotzé E., (Auth), & Senekal B. A., (Auth), AI in and for Africa: A Humanistic Perspective, Publisher: Chapman and Hall/CRC, 1st Ed., 2023. https://www.routledge.com/AI-in-and-for-Africa-A-Humanistic-Perspective/Brokensha-Kotze-Senekal/p/book/9781032231761

Artificial Intelligence in Middle East and Africa. How 112 Major Companies Benefit from AI. Outlook for 2019 and Beyond. REPORT COMMISSIONED BY MICROSOFT AND CONDUCTED BY EY. https://info.microsoft.com/rs/157-GQE-382/images/report-SRGCM1065.pdf

Artificial Intelligence for Africa: An Opportunity for Growth, Development, and Democratisation: By Access Partnership, 2018

https://accesspartnership.com/artificial-intelligence-for-africa-an-opportunity-for-growth-development-and-democratisation/

GLOSSARY OF TERMS

3D printing: A manufacturing process that creates three-dimensional objects by depositing materials layer-by-layer based on a digital model.

A

Accenture: A global professional services company that provides services in strategy, consulting, digital technology and operations.

Aerobotics (South Africa): A South African startup that uses AI and drone imagery to provide analytics to fruit farmers.

Africa Digital Financial Inclusion Facility (ADFI): A facility launched by the African Development Bank to promote digital financial inclusion across Africa.

African Development Bank (AfDB): A multilateral development finance institution that promotes economic growth and social progress across Africa.

African Digital Rights Hub: An organization that aims to advance digital rights in Africa through research, advocacy and collaboration.

African Digital Rights Network: A network of individuals and organizations promoting digital rights and inclusion in Africa.

African Digital Single Market (DSM): A proposed unified digital market across Africa to boost trade and innovation.

African Research Universities Alliance (ARUA): An alliance of African universities to strengthen research capacity and support development.

African Robotics Network (AFRON): A network that brings together robotics researchers, students and entrepreneurs across Africa.

Africar Group: A pan-African company providing ride-hailing, car rental and logistics technology services.

AI Governance: The laws, regulations, practices and standards that guide and govern the development and use of AI systems.

Alexa: A virtual assistant AI technology created by Amazon and used in smart speakers and other devices.

Algorithms: Sets of instructions or rules used by computers to solve problems and make decisions.

Amazon: A multinational technology company known for e-commerce, cloud computing and AI products.

Andela: A company that trains software engineers in Africa and connects them to jobs with global companies.

Apple's Siri: An AI-powered virtual assistant created by Apple for its devices.

Applied Law & Technology (Pty) Ltd (ALT Advisory): A South African legal and regulatory advisory firm specializing in technology.

Artificial Intelligence: Computer systems able to perform tasks normally requiring human intelligence, such as visual perception, speech recognition, and decision-making.

B

Big Data: Extremely large datasets that can be analyzed computationally to reveal patterns, trends, and associations.

Bing Chat: An AI chatbot created by Microsoft to have natural conversations and answer questions.

bioMérieux Foundation: A non-profit organization focused on increasing infectious disease diagnostics in low-resource countries.

Blockchain: A distributed database technology that permanently records transactions in a way that cannot be altered retroactively.

Brookings Institute: A nonprofit public policy organization that conducts research on major issues worldwide.

Business development: Strategies and processes for developing new business opportunities and partnerships.

C

C/C++: Common programming languages used for systems programming and desktop and web applications.

Calculus: A branch of mathematics focused on rates of change and motion between variables. Essential foundation for machine learning.

Case study: Detailed analysis of a person, group, or situation over time to study a phenomenon in context.

Chat-GPT: A conversational AI system created by Anthropic to interact in natural language.

Chatham House: An international affairs think tank based in London.

Classification: A machine learning technique that identifies which categories or classes input data belongs to.

Claude ai: A natural language AI assistant created by Anthropic to have helpful, harmless, and honest conversations.

Cloud computing: Internet-based computing services offering storage, databases, analytics, software and more on demand. Enables access to AI services.

Clustering: A machine learning technique that groups data points with similar properties into clusters. Used for exploratory analysis.

Coding: Writing computer programming code to create software programs and applications. Essential skill for AI development.

D

Data analysis: Techniques used to inspect, clean, transform and model data to discover patterns, draw conclusions, and support decision-making.

Data Bias: Systematic inaccuracies in data that lead to unfair prejudice for certain groups. Causes models to discriminate.

Data integration: Combining data from different sources and providing users a unified view of the data.

Data Management: Processes and governance for ingesting, storing, securing, preparing, and analyzing data.

Data Protection: Safeguarding data privacy rights and ensuring proper data handling and security practices.

Data Rights: Legal rights afforded to individuals regarding their personal data like consent, access, correction, and restriction of processing.

Data sovereignty: The concept that data is subject to the laws and governance of the nation it is collected and stored in.

Data storage: Computing infrastructure, devices and databases used to store and retrieve digital data.

Dataiku: A software company providing a collaborative data science platform.

Datasets: Structured collections of data used to train AI models. Quality datasets are vital for model development.

Demographics: Statistical characteristics of a population such as age, gender, income, race, and occupation. Helpful for market analysis.

Digital Transformation: Business and organizational changes accelerated by adopting digital technologies like cloud, mobile, AI and IoT.

E

E-commerce: Buying and selling goods and services over the internet. Enables access to larger markets.

Ecosystems: Interconnected networks of entities like companies, technologies, regulations, and consumers that enable innovation and growth.

Efficiency: Ability to avoid wasting resources like time, effort, energy or cost in operations and processes. AI can drive efficiency gains.

Entrepreneurship: Process of designing, launching and operating new businesses, often involving risk and innovation.

Environmental, Social and Governance (ESG): Criteria measuring a company's impact and operations across environmental, social, and governance dimensions.

Ethical: Morally correct principles and standards of conduct for individuals and organizations. Essential for responsible AI.

F

Facebook: A social media platform company that also develops AI capabilities.

Fintech: Technology innovations providing new financial services, from mobile payments to trading algorithms.

Fourth Industrial Revolution: The current era of rapid technological advances fusing physical, digital and biological worlds.

G

Generative Artificial Intelligence: AI techniques that create new content like text, images, video, and audio "from scratch" rather than classifying existing data.

Genetic engineering: Techniques for modifying or transferring genetic material in organisms to produce desired traits or results.

Google: An American multinational technology company specializing in Internet related services and products powered by innovations in AI.

Gordon Institute of Business Science (GIBS): A leading business school in South Africa.

GPS systems: Global positioning systems providing geographic location and time information anywhere on Earth. Enable mobile services and logistics.

Gross National Income (GNI) per capita: A measure of a country's economic output that accounts for GDP and income from abroad. Indicates standard of living.

H

Healthcare: Prevention, diagnosis, treatment and management of illness and the preservation of physical and mental well-being.

Human Development Index (HDI): A United Nations measure of key dimensions of human development, including life expectancy, education, and income indices.

I

Information technology: Computer systems storing, retrieving, transmitting and manipulating data, including hardware, software, databases and networking.

Innovation: Introducing new solutions through more effective concepts, products, processes, services, technologies or business models.

Insight: Deep intuitive understanding, knowledge or perception about a situation or subject.

Intel: American technology company that manufactures computer processors and develops computing innovations.

Internet of Things: Network of physical objects embedded with sensors, software and connectivity to collect and exchange data.

Investment pitch: A presentation designed to persuade potential investors to provide startup funding.

Investment: Allocating money or resources into a project or enterprise to generate profitable returns.

J

Java: A popular general purpose programming language used for desktop, web, mobile and backend applications.

Jumia: An African e-commerce company sometimes dubbed the "Amazon of Africa".

JUMO: A fintech company using AI and mobile technology to increase financial inclusion across Africa.

Keras: An open source neural network library written in Python widely used for AI development and research.

L

Linear algebra: Branch of mathematics dealing with linear equations and their transformation properties. Foundational for machine learning algorithms.

M

Machine Learning: AI techniques where systems learn and improve at tasks from data without explicit programming.

Malabo Convention: An African Union convention establishing a legal framework governing cybersecurity and personal data protection.

Marginalize: Treat a group of people as insignificant or peripheral. AI biases can marginalize certain groups.

Master Data Management System: Consolidates, maintains and provides reliable master data assets across an organization.

Mastercard: Multinational financial services corporation providing payment processing and technology services.

MDaaS Global (Nigeria): Nigerian healthtech startup using AI for diagnostics and telemedicine.

Meningitis belt: Region across sub-Saharan Africa prone to meningitis outbreaks.

Mentorship: Guidance provided by an experienced person to support professional growth and development. Valuable for entrepreneurs.

Micro-entrepreneurs: Entrepreneurs running small or informal businesses, especially in developing economies.

Microsoft: American technology company developing software, services, devices and solutions including AI capabilities.

M-Kopa: A fintech company providing pay-as-you-go solar power financing to off-grid homes across Africa.

ML Libraries: Reusable packages of machine learning code for common tasks like preprocessing, visualization and model training. Accelerate development.

mPharma: Ghanaian healthtech startup using AI and mobile technology to make medicines more accessible.

N

Netflix: Global internet entertainment service using AI for personalized recommendations and other functions.

Network of African Data Protection Authorities (NADPA): Principle encouraging African nations to establish data protection laws and authorities.

O

Orange: French multinational telecommunications corporation operating across Africa.

P

Parsons School of Design: Leading art and design school located in New York.

Pasteur Institute: Biomedical research institute with facilities across Africa researching infectious diseases.

Perfect storm: An expression meaning a rare combination of circumstances creating an event of unusual magnitude.

Policy framework: High-level plans and principles formulated to reach policy goals and guide future policy making.

Predictive analysis: is the use of statistical and machine learning techniques to analyze current and historical data to make predictions about future outcomes and trends. It provides actionable insights that can drive planning, decision-making, and predictive positioning for organizations.

Private sector: Businesses and commercial organizations not under direct government control. Drive innovation through competition.

Process automation: Using technology like AI and robotics to automate business processes requiring manual effort, to improve efficiency.

Processing power: Speed at which a computer can execute instructions and process data. AI needs high processing power, often with specialized hardware.

Product design: Creating the form, features and function of a new product to solve customer problems. Focus on human-cantered design.

Product development: Entire process of bringing a new product from concept to market launch including design, engineering, testing and marketing.

Project management: Planning, organizing and managing tasks and resources to achieve project objectives on time and on budget. Crucial ability for startups.

Prototype: Early sample model of a product built to demonstrate concepts and test functionality. Allows gathering user feedback.

Python: A widely used high-level programming language known for code readability. Extensively used for AI programming and data science.

PyTorch: An open source machine learning library for Python based on tensor computation like NumPy, with strong GPU support.

Q

Quantum computing: Next-generation computing that harnesses quantum mechanical phenomena for exponential speedups in processing power.

R

R: A programming language specialized for statistical analysis and graphical visualization. Commonly used for data science.

Regression: A machine learning technique used to estimate relationships between variables for predictive analysis.

Risk management: Identifying, evaluating and mitigating any uncertainties that could threaten the success of activities and projects.

Robotics: Field dealing with the design, construction, operation and use of robots. Automates manual repetitive tasks.

S

Self-driving cars: Autonomous vehicles sensing surroundings and navigating without human input using AI and sensor technology.

Siri: See Apple's Siri.

Soft Skills: Personal attributes and interpersonal skills like teamwork, communication and emotional intelligence that enable workplace success.

Stakeholders: People and organizations impacted by and influencing a project or entity's outcomes.

Startups: Newly formed businesses working to develop scalable and innovative business models.

Statistics: Mathematical techniques using quantified data analysis to make inferences, discoveries and informed decisions. Core component of data science.

Swvl (Egypt): Egyptian mass transit startup providing AI optimized public bus routes based on user demand.

T

Telecommunications infrastructure: Physical facilities supporting communications including cable networks, telephone lines, cellular towers and satellites. Enables digital services.

Telemedicine: Remote healthcare services like health monitoring, consultations and diagnoses enabled by telecommunications and AI technologies.

TensorFlow: End-to-end open source machine learning platform created by Google for training, deploying and productionizing AI models.

The African Institute for Mathematical Sciences (AIMS): A pan-African network of centres of excellence for postgraduate training, research, and outreach in mathematical sciences.

The African Union: A union consisting of the 55 member states that make up the countries of the African Continent.

The Council for Scientific and Industrial Research (CSIR): South Africa's central scientific research and development organization.

The National Research Foundation: Supports and promotes research and innovation in South Africa.

U

UNICEF: United Nations organization providing humanitarian and developmental aid to children worldwide.

United Nations (UN): Intergovernmental organization working to maintain international peace and security, protect human rights, deliver humanitarian aid, and promote sustainable development.

United Nations Development Programme (UNDP): UN organization eradicating poverty, reducing inequalities and building resilience through sustainable development.

University of Pretoria: Leading research university located in South Africa.

V

Verisk Maplecroft: Global risk analytics company assessing threats to business operations worldwide.

Virtual Centre: Proposed single Africa wide e-commerce platform to boost intra-African trade. (Also distributed customer support with agents in remote offices or working from home).

Visualization: Graphical representation of data and patterns used to intuitively discern insights from complex information.

Voyc.ai (South Africa): South African startup offering an AI assistant for retail customer engagement.

W

Web3: Evolving concept of the internet built around decentralized blockchain technology and tokenized economics.

West Africa Data Protection and Privacy Rights Coalition (WADPPRC): Coalition promoting data privacy in West Africa through advocacy.

World Economic Forum: International organization engaging public and private stakeholders to shape global systems.

Y

YouTube: Online video sharing platform owned by Google with over 2 billion monthly users.